Problem Books in Mathematics

Edited by K.A. Bencsáth
P.R. Halmos

AF651673

Edward J. Barbeau

Pell's Equation

Springer

Edward J. Barbeau
Department of Mathematics
University of Toronto
100 St. George Street
Toronto, Ontario M5S 3G3
Canada
barbeau@math.toronto.edu

Series Editor
Katalin A. Benscáth
Mathematics
School of Science
Manhattan College
Riverdale, NY 10471
USA
katalin.benscath@manhattan.edu

Paul R. Halmos
Department of Mathematics
Santa Clara University
Santa Clara, Ca 95053
USA
phalmos@scuacc.scu.edu

With 9 illustrations.

Mathematics Subject Classification (2000): 26-06, 26Bxx, 34Axx, 30Axx

Library of Congress Cataloging-in-Publication Data
Barbeau, Edward, 1938-
Pell's equation / Edward J. Barbeau.
p.cm. –(Problem books in mathematics)
Includes bibliographical references and index.
ISBN 0-387-95529-1
1. Pell's equation. 2. Number theory. I. Title. II. Series.
QA242 .B4395 2002
512'.7—dc21 2002070735

ISBN 0-387-95529-1

All rights reserved. This work may not be translated or copied in whole or in part without the written permission of the publisher (Springer Science + Business Media, LLC, 233 Spring Street, New York, NY 10013, USA), except for brief excerpts with reviews or scholarly analysis. Use in connection with any form of information storage and retrieval, electronic adaptation, computer software, or by similar or dissimilar methodology now known or hereafter developed is forbidden.
The use in this publication of trade names, trademarks, service marks, and similar terms, even if they are not identified as such, is not to be taken as an expression of opinion as to whether or not they are subject to propietary rights.

springer.com

Barbeau: Pell's Equation

© 2003 Springer-Verlag New York, Inc.

All rights reserved. No part of this publication may be reproduced, stored in any electronic or mechanical form, including photocopy, recording or otherwise, without the prior written permission of the publisher.

First Indian Reprint 2011

ISBN 978-81-8489-851-4

"This edition is licensed for sale only in India, Pakistan, Bangladesh, Sri Lanka and Nepal. Circulation of this edition outside of these territories is UNAUTHORIZED AND STRICTLY PROHIBITED."

This edition is published by Springer (India) Private Limited, (a part of Springer Science+Business Media), Registered Office: 3rd Floor, Gandharva Mahavidyalaya, 212, Deen Dayal Upadhyaya Marg, New Delhi – 110 002, India

Printed in India by Rashtriya Printers, Delhi.

To my grandchildren
Alexander Joseph Gargaro
Maxwell Edward Gargaro
Victoria Isabelle Barbeau
Benjamin Maurice Barbeau

Preface

This is a focused exercise book in algebra.

Facility in algebra is important for any student who wants to study advanced mathematics or science. An algebraic expression is a carrier of information. Sometimes it is easy to extract the information from the form of the expression; sometimes the information is latent, and the expression has to be altered to yield it up. Thus, students must learn to manipulate algebraic expressions judiciously with a sense of strategy. This sense of working towards a goal is lacking in many textbook exercises, so that students fail to gain a sense of the coherence of mathematics and so find it difficult if not impossible to acquire any significant degree of skill.

Pell's equation seems to be an ideal topic to lead college students, as well as some talented and motivated high school students, to a better appreciation of the power of mathematical technique. The history of this equation is long and circuitous. It involved a number of different approaches before a definitive theory was found. Numbers have fascinated people in various parts of the world over many centuries. Many puzzles involving numbers lead naturally to a quadratic Diophantine equation (an algebraic equation of degree 2 with integer coefficients for which solutions in integers are sought), particularly ones of the form $x^2 - dy^2 = k$, where d and k are integer parameters with d nonsquare and positive. A few of these appear in Chapter 2. For about a thousand years, mathematicians had various ad hoc methods of solving such equations, and it slowly became clear that the equation $x^2 - dy^2 = 1$ should always have positive integer solutions other than $(x, y) = (1, 0)$. There were some partial patterns and some quite effective methods of finding solutions, but a complete theory did not emerge until the end of the eighteenth century. It is unfortunate that the equation is named after a seventeenth-century English mathematician, John Pell, who, as far as anyone can tell, had hardly anything to do with it. By his time, a great deal of spadework had been done by many Western European mathematicians. However, Leonhard Euler, the foremost European mathematician of the eighteenth century, who *did* pay a lot of attention to the equation, referred to it as "Pell's equation" and the name stuck.

In the first three chapters of the book the reader is invited to explore the situation, come up with some personal methods, and then match wits with early Indian and

European mathematicians. While these investigators were pretty adept at arithmetic computations, you might want to keep a pocket calculator handy, because sometimes the numbers involved get pretty big. Just try to solve $x^2 - 61y^2 = 1$! So far there is not a clean theory for the higher-degree analogues of Pell's equation, although a great deal of work was done on the cubic equation by such investigators as A. Cayley, P.H. Daus, G.B. Mathews, and E.S. Selmer in the late nineteenth and early twentieth century; the continued fraction technique seems to be so special to the quadratic case that it is hard to see what a proper generalization might be. As sometimes happens in mathematics, the detailed study of particular cases becomes less important and research becomes more focused on general structure and broader questions. Thus, in the last fifty years, the emphasis has been on the properties of larger classes of Diophantine equations. Even the resolution of the Fermat Conjecture, which dealt with a particular type of Diophantine equation, by Andrew Wiles was done in the context of a very broad and deep study. However, this should not stop students from going back and looking at particular cases. Just because professional astronomers have gone on to investigating distant galaxies and seeking knowledge on the evolution of the universe does not mean that the backyard amateur might not find something of interest and value about the solar system.

The subject of this book is not a mathematical backwater. As a recent paper of H.W. Lenstra in the *Notices of the American Mathematical Society* and a survey paper given by H.C. Williams at the Millennial Conference on Number Theory in 2000 indicate, the efficient generation of solutions of an ordinary Pell's equation is a live area of research in computer science. Williams mentions that over 100 articles on the equation have appeared in the 1990s and draws attention to interest on the part of cryptographers. Pell's equation is part of a central area of algebraic number theory that treats quadratic forms and the structure of the rings of integers in algebraic number fields. Even at the specific level of quadratic Diophantine equations, there are unsolved problems, and the higher-degree analogues of Pell's equation, particularly beyond the third, do not appear to have been well studied. This is where the reader might make some progress.

The topic is motivated and developed through sections of exercises that will allow the student to recreate known theory and provide a focus for algebraic practice. There are several *explorations* that encourage the reader to embark on individual research. Some of these are numerical, and often require the use of a calculator or computer. Others introduce relevant theory that can be followed up on elsewhere, or suggest problems that the reader may wish to pursue.

The opening chapter uses the approximations to the square root of 2 to indicate a context for Pell's equation and introduce some key ideas of recursions, matrices, and continued fractions that will play a role in the book. The goal of the second chapter is to indicate problems that lead to a Pell's equation and to suggest how mathematicians approached solving Pell's equation in the past. Three chapters then cover the core theory of Pell's equation, while the sixth chapter digresses to draw out some connections with Pythagorean triples. Two chapters embark on the study of higher-degree analogues of Pell's equation, with a great deal left to the reader to pursue. Finally, we look at Pell's equation modulo a natural number.

I have used some of the material of this book in a fourth-year undergraduate research seminar, as well as with talented high school students. It has also been the basis of workshops with secondary teachers. A high school background in mathematics is all that is needed to get into this book, and teachers and others interested in mathematics who do not have (or have forgotten) a background in advanced mathematics may find that it is a suitable vehicle for keeping up an independent interest in the subject. Teachers could use it as a source of material for their more able students.

There are nine chapters, each subdivided into sections. Within the same chapter, Exercise z in Section y is referred to as *Exercise* $y.z$; if reference is made to an exercise in a different chapter x, it will be referred to as *Exercise* $x.y.z$. The end of an exercise may be indicated by ♠ to distinguish it from explanatory text that follows. Within each chapter there are a number of *Explorations*; these are designed to raise other questions that are in some way connected with the material of the exercises. Some of the explanations may be thought about, and then returned to later when the reader has worked through more of the exercises, since occasionally later work may shed additional light. It is hoped that these explorations may encourage students to delve further into number theory. A glossary of terms appears at the end of the book.

I would like to thank anonymous reviewers for some useful comments and references, a number of high school and undergraduate university students for serving as guinea pigs for some of the material, and my wife, Eileen, for her support and patience.

Contents

1

The Square Root of 2

To arouse interest in Pell's equation and introduce some of the ideas that will be important in its study, we will examine the question of the irrationality of the square root of 2. This has its roots in Greek mathematics and can be looked at from the standpoint of arithmetic, geometry, or analysis.

The standard arithmetical argument for the irrationality of 2 goes like this. Suppose that $\sqrt{2}$ is rational. Then we can write it as a fraction whose numerator and denominator are positive integers. Let this fraction be written in lowest terms: $\sqrt{2} = p/q$, with the greatest common divisor of p and q equal to 1. Then $p^2 = 2q^2$, so that p must be even. But then p^2 is divisible by 4, which means that q must be even along with p. Since this contradicts our assumption that the fraction is in lowest terms, we must abandon our supposition that the square root of 2 is rational.

Another way of looking at this argument is to note that if $\sqrt{2} = p/q$ as above, then, since numerator and denominator are both even, we can write it as a fraction with strictly smaller numerator and denominator, and can continue doing this indefinitely. This is impossible. This "descent" approach has an echo in a geometric argument given below. This will, in turn, bring into play the role of recursions.

1.1 Can the Square Root of 2 Be Rational?

Two alternative arguments, one in Exercise 1.1 and the other in Exercises 1.2–1.4, are presented. The second argument introduces two sequences, defined recursively, that figure in solutions to the equation $x^2 - 2y^2 = \pm 1$.

Exercise 1.1. An attractive argument that $\sqrt{2}$ is not a rational number utilizes the geometry of the square. Let $ABCD$ be a square with diagonal AC.

(a) Determine a point E on AC for which $AE = BC$. Let F be a point selected on BC such that $EF \perp AC$. Prove that $BF = FE$.

(b) Complete the square $CEFG$. Suppose that the lengths of the side and diagonal of square $ABCD$ are a_1 and b_1, respectively. Prove that the lengths of the side and diagonal of the square $CEFG$ are $b_1 - a_1$ and $2a_1 - b_1$, respectively.

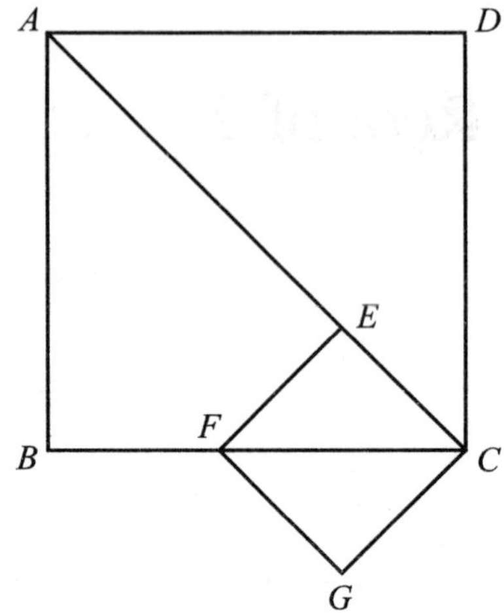

FIGURE 1.1.

(c) We can perform the same construction on square $CEFG$ to produce an even smaller square and so continue indefinitely. Suppose that the sides of the successive squares have lengths $a_1, a_2, a_3, \ldots$, and their corresponding diagonals have lengths $b_1, b_2, b_3, \ldots$. Argue that both sequences are decreasing and that they jointly satisfy the recursion relations

$$a_n = b_{n-1} - a_{n-1},$$
$$b_n = 2a_{n-1} - b_{n-1},$$

for $n \geq 1$.

(d) Suppose that the ratio b_1/a_1 of the diagonal and side lengths of the square $ABCD$ is equal to the ratio p/q of positive integers p and q. This means that there is a length λ for which $a_1 = q\lambda$ and $b_1 = p\lambda$ (or as the Greeks might have put it, the length λ "measures" both the side and diagonal of the square). Verify that $a_2 = (p - q)\lambda$ and $b_2 = (2q - p)\lambda$.

(e) Prove that for each positive integer n, both a_n and b_n are positive integer multiples of λ.

(f) Arithmetically, observe that there are only finitely many pairs smaller than (p, q) that can serve as multipliers of λ when we construct smaller squares as described. Deduce that only finitely many squares are thus constructible.

Geometrically, note that we can repeat the process as often as desired. Now complete the contradiction argument and deduce that the ratio of the lengths of the diagonal and side of a square is not rational. ♠

We can pick up the recursion theme in another way, in this case getting a sequence of pairs of integers that increase rather than decrease in size and whose ratio will approximate $\sqrt{2}$ more and more closely. We begin with the simple observation that $(\sqrt{2}-1)(\sqrt{2}+1) = 1$. This will give us an equation involving the square root of 2 that will form the basis of a recursion.

Exercise 1.2.

(a) Verify that

$$\sqrt{2} = 1 + \frac{1}{1+\sqrt{2}}.$$

(b) Replace the $\sqrt{2}$ on the right side by the whole of the right side to obtain

$$\sqrt{2} = 1 + \frac{1}{1+1+\frac{1}{1+\sqrt{2}}} = 1 + \frac{1}{2+\frac{1}{1+\sqrt{2}}}. \quad \spadesuit$$

For convenience, we will write this as

$$\sqrt{2} = 1 + 1/2 + 1/1 + \sqrt{2},$$

where each slash embraces all of what follows it as the denominator of a fraction.

On the basis of Exercise 1.2(b), we are tempted to write an infinite *continued fraction* representation

$$\sqrt{2} = 1 + 1/2 + 1/2 + 1/2 + 1/2 + \cdots.$$

This can be justified by defining the infinite representation as the limit of a sequence, as we shall see in the next exercise.

Exercise 1.3.

(a) Let $r_1 = 1$ and define recursively, for each integer $n = 2, 3, 4, \ldots,$

$$r_n = 1 + \frac{1}{1+r_{n-1}}.$$

Write out the first ten terms of this sequence. Observe that for each positive integer n, $r_n = p_n/q_n$, where p_n and q_n are coprime integers. Construct a table, listing beside each index n the values of r_n as both a common and a decimal fraction, as well as the values of p_n and q_n. Look for patterns. Is the sequence $\{r_n\}$ increasing? decreasing? What happens to r_n as n gets larger? (You should verify that each r_n has a terminating representation of the form $1 + 1/2 + 1/2 + 1/2 + \cdots + 1/2 + 1/2$.)

(b) Verify that for each positive integer n,

$$r_{n+1} - r_n = -\frac{(r_n - r_{n-1})}{(1+r_n)(1+r_{n-1})}.$$

(c) Prove that for each pair of positive integers k and l,

$$1 \le r_1 \le r_3 \le \cdots \le r_{2k+1} \le \cdots \le r_{2l} \le \cdots \le r_4 \le r_2 \le \frac{3}{2}$$

and

$$|r_{n+1} - r_n| \leq \frac{1}{4}|r_n - r_{n-1}|.$$

(d) From (c), we find that as n increases, r_n oscillates around and gets closer to a limiting value α that lies between 1 and $\frac{3}{2}$. Argue that

$$\alpha = 1 + \frac{1}{1+\alpha}$$

and deduce that $\alpha^2 = 2$, so that $\alpha = \sqrt{2}$. ♠

For each n, the number r_n can be represented as a finite continued fraction that is the beginning of the representation for $\sqrt{2}$ obtained in Exercise 1.2. Since the limit of the sequence $\{r_n\}$ is $\sqrt{2}$, the infinite continued fraction representing $\sqrt{2}$ is the limit of finite ones of increasing length.

Exercise 1.4. The representation of $\sqrt{2}$ as a continued fraction suggests an alternative method of verifying that $\sqrt{2}$ is not equal to a common fraction p/q.

(a) Let s be an arbitrary positive real number and let

$$a_0 = \lfloor s \rfloor$$

represent the largest integer that does not exceed s. (This is called the "floor" of s.) Then

$$s = a_0 + b_1,$$

where $0 \leq b_1 < 1$. If $b_1 > 0$, let $s_1 = 1/b_1$, so that $s_1 > 1$. Verify that

$$s = a_0 + \frac{1}{s_1}.$$

Show that s_1 is either an integer or can be written in the form

$$a_1 + \frac{1}{s_2},$$

where $a_1 = \lfloor s_1 \rfloor$ and $s_2 > 1$. Thus, verify that

$$s = a_0 + 1/a_1 + 1/s_2.$$

This process can be continued. At the nth stage, suppose that $a_0, a_1, \ldots, a_{n-1}, s_n$ have been chosen such that

$$s = a_0 + 1/a_1 + 1/a_2 + 1/a_3 + \cdots + 1/a_{n-1} + 1/s_n.$$

If s_n is an integer, the process terminates. Otherwise, let $a_n = \lfloor s_n \rfloor$ and let $s_{n+1} > 1$ satisfy

$$s_n = a_n + \frac{1}{s_{n+1}}.$$

If s_n is not an integer, the process continues.

(b) Verify that when $s = 17/5$, the process just described yields $3 + 1/2 + 1/2$.
(c) Carry out the process when $s = 347/19$.
(d) Observe that $\lfloor\sqrt{2}\rfloor = 1$. Carry out the process when $s = \sqrt{2}$. Does it terminate? Why?
(e) Let $s = p_0/p_1$ be a rational number with p_0 and p_1 coprime positive integers. Write

$$\frac{p_0}{p_1} = a_0 + \frac{p_2}{p_1},$$

where a_0 is a nonnegative integer and where $0 \leq p_2 < p_1$. If $p_2 \neq 0$, we can similarly write

$$\frac{p_1}{p_2} = a_1 + \frac{p_3}{p_2}$$

with $0 \leq p_3 < p_2$. Suppose that s is developed as a continued fraction as in (a). Show that for $n \geq 1$, either $p_{n+1} = 0$ or s_n can be written in the form p_n/p_{n+1}, where

$$p_{n-1} = a_{n-1}p_n + p_{n+1}$$

and $0 < p_{n+1} < p_n$. Deduce from this that the continued fraction process must terminate.
(f) Deduce from (d) and (e) that $\sqrt{2}$ is not rational.

Exercise 1.5. Let $\{p_n\}$, $\{q_n\}$, and $\{r_n\}$ be the sequences defined in Exercise 1.3.
(a) Prove that for $n \geq 2$,

$$p_n = p_{n-1} + 2q_{n-1},$$
$$q_n = p_{n-1} + q_{n-1}.$$

(b) Prove, by induction, that

$$p_n^2 - 2q_n^2 = (-1)^n,$$

so that

$$r_n^2 - 2 = \frac{(-1)^n}{q_n^2}$$

and

$$r_n - \sqrt{2} = \frac{(-1)^n}{q_n^2(r_n + \sqrt{2})}.$$

(c) Use (b) to show that as n gets larger and larger, r_n gets closer and closer to $\sqrt{2}$.

Exercise 1.6. Previous exercises introduced two sets of recursions,

$$a_n = b_{n-1} - a_{n-1}; \quad b_n = 2a_{n-1} - b_{n-1},$$

and

$$p_n = p_{n-1} + 2q_{n-1}; \quad q_n = p_{n-1} + q_{n-1}.$$

The first recursion produced decreasing pairs of positive real numbers, while the second produced increasing pairs. Verify that

$$b_{n-1} = b_n + 2a_n \quad \text{and} \quad a_{n-1} = b_n + a_n,$$

and

$$q_{n-1} = p_n - q_n \quad \text{and} \quad p_{n-1} = 2q_n - p_n,$$

so that each recursion is the inverse of the other in the sense that the recursion relation for each gives the recursion relation for the other for decreasing rather than increasing values of the index n.

Exercise 1.7. From Exercise 1.5, we see that $(x, y) = (p_{2k-1}, q_{2k-1})$ satisfies the equation $x^2 - 2y^2 = -1$, while $(x, y) = (p_{2k}, q_{2k})$ satisfies the equation $x^2 - 2y^2 = 1$.

(a) Do you think that these constitute a *complete* set of solutions to the two equations in positive integers x and y? Why?
(b) Argue that for each index n, p_n is odd, while q_n has the same parity as n.
(c) Use the result of (b) to find solutions in positive integers x and y to $x^2 - 8y^2 = 1$.
(d) Are there any solutions in positive integers to $x^2 - 8y^2 = -1$?

Exploration 1.1. Many people are attracted to mathematics by its fecundity. In particular, the sequences $\{p_n\}$ and $\{q_n\}$ have a richness that is fun to investigate.

Consider the table

n	p_n	q_n	p_nq_n
0	1	0	0
1	1	1	1
2	3	2	6
3	7	5	35
4	17	12	204
5	41	29	1189
6	99	70	6930
...	...	...	...

where $p_n^2 - 2q_n^2 = (-1)^n$. Using and extending this table, one can find an abundance of patterns. Describe as many of these as you can and try to prove whether they hold in general. In Exercise 1.5 we found that each of the sequences $\{p_n\}$ and $\{q_n\}$ can be determined recursively by relations that involve *both* sequences. Try to establish recursions for each sequence that involve only its own earlier entries and not those of the other ones. You may find that discovering the patterns is more difficult than proving them; in many cases, an induction argument will do the job. Use these patterns to extend the table further; check your results by computing $p_n^2 - 2q_n^2$.

Here is a rather interesting pattern for you to describe and try to prove a generalization for:

$$\begin{aligned} 3^4 - 5 \times 4^2 &= 1, \\ 7^4 - 24 \times 10^2 &= 1, \\ 17^4 - 145 \times 24^2 &= 1, \\ 41^4 - 840 \times 58^2 &= 1. \end{aligned}$$

Exploration 1.2. What are the possible values that $x^2 - 2y^2$ can assume when x and y are integers? In particular, can $x^2 - 2y^2$ be equal to 2? −2? 3? −3?

1.2 A Little Matrix Theory

We can think of the sequences $\{p_n\}$ and $\{q_n\}$ of the first section as being paired, so that there is an operation that acts upon them jointly that allows us to derive each pair (p_n, q_n) from its predecessor (p_{n-1}, q_{n-1}). This operation involves the coefficients 1, 2; 1, 1 that occur in the relations $p_n = p_{n-1} + 2q_{n-1}$ and $q_n = p_{n-1} + q_{n-1}$.

A convenient way to encode this operation is through a display of the coefficients in a square array. Indeed, the entities we shall define are more than merely a vehicle for coding; they permit the definition of algebraic manipulations that will enable us to obtain new results.

In order to represent the relation between successive pairs $\{p_n, q_n\}$ by a single equation, we introduce the concept of a matrix. An array

$$\begin{pmatrix} a & b \\ c & d \end{pmatrix}$$

is called a 2×2 matrix and $\binom{u}{v}$ a column vector. We define

$$\begin{pmatrix} a & b \\ c & d \end{pmatrix}\begin{pmatrix} u \\ v \end{pmatrix} = \begin{pmatrix} au + bv \\ cu + dv \end{pmatrix}.$$

This equation can be regarded as recording the result of performing an operation on a column vector to obtain another column vector.

The power of this method of expressing the transformation resides in our ability to impose an algebraic structure on the sets of vectors and matrices. The sum of two vectors and the product of a constant and a vector are defined coordinatewise:

$$\begin{pmatrix} u \\ v \end{pmatrix} + \begin{pmatrix} r \\ s \end{pmatrix} = \begin{pmatrix} u + r \\ v + s \end{pmatrix}$$

and

$$k\begin{pmatrix} u \\ v \end{pmatrix} = \begin{pmatrix} ku \\ kv \end{pmatrix}.$$

We can make a similar definition for the sum of two matrices and the product of a number and a matrix:

$$\begin{pmatrix} p & q \\ r & s \end{pmatrix} + \begin{pmatrix} a & b \\ c & d \end{pmatrix} = \begin{pmatrix} p+a & q+b \\ r+c & s+d \end{pmatrix}$$

and

$$k\begin{pmatrix} a & b \\ c & d \end{pmatrix} = \begin{pmatrix} ka & kb \\ kc & kd \end{pmatrix}.$$

There is nothing to stop us from performing two successive operations on a column vector, by multiplying it on the left by two matrices one after the other. As we will see in the first exercise, the result of doing this is equivalent to multiplying the vector by a single matrix. We use this fact to define the product of matrices. There are many ways to define such a product, and the one used here does not seem to be the most natural, but it is the standard one and reflects the role of matrices as entities that act upon something.

Exercise 2.1. Verify that

$$\begin{pmatrix} p & q \\ r & s \end{pmatrix}\left[\begin{pmatrix} a & b \\ c & d \end{pmatrix}\begin{pmatrix} u \\ v \end{pmatrix}\right] = \begin{pmatrix} pa+qc & pb+qd \\ ra+sc & rb+sd \end{pmatrix}\begin{pmatrix} u \\ v \end{pmatrix}. \spadesuit$$

This motivates the definition of the product of two 2×2 matrices by

$$\begin{pmatrix} p & q \\ r & s \end{pmatrix}\begin{pmatrix} a & b \\ c & d \end{pmatrix} = \begin{pmatrix} pa+qc & pb+qd \\ ra+sc & rb+sd \end{pmatrix}.$$

This definition produces that matrix that has the same effect on a vector as the application of its two multipliers one after another. As you will see in the following exercises, it is the appropriate definition to use in pursuit of recursions for the sequences $\{p_n\}$ and $\{q_n\}$, and in fact, it helps us understand why both sequences satisfy the same recursion relation.

Exercise 2.2.

(a) Verify that for $n \geq 2$,

$$\begin{pmatrix} p_n \\ q_n \end{pmatrix} = \begin{pmatrix} 1 & 2 \\ 1 & 1 \end{pmatrix}\begin{pmatrix} p_{n-1} \\ q_{n-1} \end{pmatrix}.$$

(b) Verify that

$$\begin{pmatrix} 3 & 4 \\ -2 & 1 \end{pmatrix}\begin{pmatrix} -1 & 3 \\ 5 & 2 \end{pmatrix} = \begin{pmatrix} 17 & 17 \\ 7 & -4 \end{pmatrix}$$

and that

$$\begin{pmatrix} -1 & 3 \\ 5 & 2 \end{pmatrix}\begin{pmatrix} 3 & 4 \\ -2 & 1 \end{pmatrix} = \begin{pmatrix} -9 & -1 \\ 11 & 22 \end{pmatrix}. \spadesuit$$

Exercise 2.3.

(a) Defining the square of a matrix to be the product of the matrix with itself, verify that

$$\begin{pmatrix} a & b \\ c & d \end{pmatrix}^2 = \begin{pmatrix} a^2 + bc & b(a+d) \\ c(a+d) & d^2 + bc \end{pmatrix}$$
$$= (a+d)\begin{pmatrix} a & b \\ c & d \end{pmatrix} - (ad - bc)\begin{pmatrix} 1 & 0 \\ 0 & 1 \end{pmatrix}.$$

In particular, verify that

$$\begin{pmatrix} 1 & 2 \\ 1 & 1 \end{pmatrix}^2 = 2\begin{pmatrix} 1 & 2 \\ 1 & 1 \end{pmatrix} + \begin{pmatrix} 1 & 0 \\ 0 & 1 \end{pmatrix}.$$

(b) Verify that

$$\begin{pmatrix} p_{n+1} \\ q_{n+1} \end{pmatrix} = \begin{pmatrix} 3 & 4 \\ 2 & 3 \end{pmatrix}\begin{pmatrix} p_{n-1} \\ q_{n-1} \end{pmatrix} = \begin{pmatrix} 1 & 2 \\ 1 & 1 \end{pmatrix}^2 \begin{pmatrix} p_{n-1} \\ q_{n-1} \end{pmatrix}.$$

(c) Use (b) and (c) to prove that

$$p_{n+1} = 2p_n + p_{n-1} \quad \text{and} \quad q_{n+1} = 2q_n + q_{n-1}$$

for $n \geq 1$.

Exercise 2.4.

(a) Prove that for $n \geq 1$,

$$p_n q_n = p_n q_{n-1} + p_{n-1} q_n + p_{n-1} q_{n-1}.$$

(b) Prove that for $n \geq 1$,

$$p_{n+1} q_{n+1} = 6 p_n q_n - p_{n-1} q_{n-1}. \quad ♠$$

The quadratic form $x^2 - 2y^2$ can also be expressed in terms of matrices. We have seen how to define the product of a matrix and a column vector on the right. In an analogous way, it is possible to define the product of a matrix and a row vector on the left:

$$(u, v)\begin{pmatrix} a & b \\ c & d \end{pmatrix} = (ua + vc, ub + vd).$$

The product of a row vector and a column vector is defined to be the number

$$(u_1, u_2)\begin{pmatrix} v_1 \\ v_2 \end{pmatrix} = u_1 v_1 + u_2 v_2.$$

Let

$$U = (u_1, u_2), \quad A = \begin{pmatrix} a & b \\ c & d \end{pmatrix}, \quad \text{and} \quad V = \begin{pmatrix} v_1 \\ v_2 \end{pmatrix}.$$

The threefold product UAV can be interpreted as $(UA)V$ or $U(AV)$. Is it possible that these are different?

Exercise 2.5.

(a) Verify that with the notation as defined above, $(UA)V$ and $U(AV)$ are both equal to $au_1v_1 + bu_1v_2 + cu_2v_1 + du_2v_2$.

(b) Verify that

$$(x, y)\begin{pmatrix} 1 & 0 \\ 0 & -2 \end{pmatrix}\begin{pmatrix} x \\ y \end{pmatrix} = x^2 - 2y^2.$$

Exercise 2.6.

(a) Verify that the mapping

$$\begin{pmatrix} x \\ y \end{pmatrix} \longrightarrow \begin{pmatrix} 3 & 4 \\ 2 & 3 \end{pmatrix}\begin{pmatrix} x \\ y \end{pmatrix}$$

preserves the value of the form $x^2 - 2y^2$. This means that you must show that

$$(3x + 4y)^2 - 2(2x + 3y)^2 = x^2 - 2y^2. \quad ♠$$

We can gain further understanding of this invariance of the form $x^2 - 2y^2$ through an analysis of the interrelations among the matrices involved. The mapping acting on the pair (x, y) discussed earlier can be expressed in the alternative format

$$(x, y) \longrightarrow (x, y)\begin{pmatrix} 3 & 2 \\ 4 & 3 \end{pmatrix}.$$

The matrix $\left(\begin{smallmatrix} 3 & 2 \\ 4 & 3 \end{smallmatrix}\right)$ obtained by reflecting the matrix $\left(\begin{smallmatrix} 3 & 4 \\ 2 & 3 \end{smallmatrix}\right)$ about its diagonal is called the *transpose* of the latter matrix.

Exercise 2.7. Verify that the result of Exercise 2.6 amounts to the assertion that

$$(x, y)\begin{pmatrix} 3 & 2 \\ 4 & 3 \end{pmatrix}\begin{pmatrix} 1 & 0 \\ 0 & -2 \end{pmatrix}\begin{pmatrix} 3 & 4 \\ 2 & 3 \end{pmatrix}\begin{pmatrix} x \\ y \end{pmatrix} = (x, y)\begin{pmatrix} 1 & 0 \\ 0 & -2 \end{pmatrix}\begin{pmatrix} x \\ y \end{pmatrix}$$

and check that indeed

$$\begin{pmatrix} 3 & 2 \\ 4 & 3 \end{pmatrix}\begin{pmatrix} 1 & 0 \\ 0 & -2 \end{pmatrix}\begin{pmatrix} 3 & 4 \\ 2 & 3 \end{pmatrix} = \begin{pmatrix} 1 & 0 \\ 0 & -2 \end{pmatrix}.$$

1.3 Pythagorean Triples

According to Pythagoras's theorem, the area of the square raised on the hypotenuse of a right triangle is equal to the sum of the areas of the squares inscribed on the legs (other two sides). If the lengths of the legs and hypotenuse are, respectively, a, b, c, then

$$a^2 + b^2 = c^2.$$

When a, b, c are integers, we say that (a, b, c) is a *Pythagorean triple*. Many readers will be familiar with the triples (3, 4, 5), (5, 12, 13), and (8, 15, 17). What other ones are there?

It is not possible to have Pythagorean triples with the smallest two numbers equal. For this would correspond to isosceles right triangles for which the length of the hypotenuse is $\sqrt{2}$ times the length of either arm, and we have seen that $\sqrt{2}$ is not equal to the ratio of two integers. However, we can come pretty close in that it *is* possible for the smallest two numbers of a Pythagorean triple to differ by only 1. One familiar example is the triple (3, 4, 5), but there are infinitely many others. Before investigating this, we will review the parametric formula that gives all possible Pythagorean triples.

Exercise 3.1.

(a) Prove that for any integers k, m, n,

$$(k(m^2 - n^2), 2kmn, k(m^2 + n^2))$$

is a Pythagorean triple.

(b) Suppose that (a, b, c) is a Pythagorean triple with b even. Is it possible to find integers k, m, n for which

$$a = k(m^2 - n^2), \quad b = 2kmn, \quad c = k(m^2 + n^2)?$$

Experiment with specific examples.

Exercise 3.2.

(a) Consider the case of Pythagorean triples (a, b, c) in which the two smallest entries differ by 1. These entries must be of the form $m^2 - n^2$ and $2mn$ with difference equal to $+1$ or -1. Derive, for each case, a condition of the form $x^2 - 2y^2 = 1$ where x and y are dependent on m and n.

(b) $(x, y) = (3, 2)$ satisfies $x^2 - 2y^2 = 1$. Determine from this equation two possible corresponding values of the pair (m, n) and their Pythagorean triples.

(c) Use solutions in integers for $x^2 - 2y^2 = 1$ to obtain other Pythagorean triples (a, b, c) with $b = a + 1$, and look for patterns. We will explore some of these in Exercise 3.4.

Exercise 3.3. Franz Gnaédinger in Zurich has given an interesting method of generating triples $(a, a + 1, c)$ for which either $a^2 + (a + 1)^2 = c^2$ or $a^2 + (a + 1)^2 = c^2 + 1$. Begin with the triple (0, 1, 0) and repeatedly apply the transformation

$$(a, a + 1, c) \longrightarrow (a + c, a + c + 1, 2a + c + 1).$$

(a) Verify that $(0, 1, 0) \to (0, 1, 1) \to (1, 2, 2) \to (3, 4, 5) \to (8, 9, 12) \to (20, 21, 29)$.

(b) Prove that one obtains solutions to $x^2 + y^2 = z^2$ and $x^2 + y^2 = z^2 + 1$ alternately.

Exercise 3.4. Consider the sequence 0, 1, 2, 5, 12, 29, 70, ... This is the sequence $\{q_n\}$ given in Section 1. Recall that $q_{n+1} = 2q_n + q_{n-1}$ for $n \geq 2$.

(a) Show that if we take n and m to be two consecutive terms in this sequence, then

$$(m^2 - n^2, 2mn, m^2 + n^2)$$

is a Pythagorean triple whose smallest entries differ by 1.

(b) Verify that this process yields the following list of parametric pairs with their corresponding Pythagorean triples. Extend the list further.

(m, n)	(a, b, c)
(2, 1)	(3, 4, 5)
(5, 2)	(21, 20, 29)
(12, 5)	(119, 120, 169)
(29, 12)	(697, 696, 985)

(c) Justify the following rule for finding Pythagorean triples of the required type: Form the sequence $\{1, 6, 35, 204, 1189, \ldots\}$. It starts with the numbers 1 and 6. Subsequent terms are obtained as follows: Let u and v be two consecutive terms in this order; the next term is $6v - u$. The largest number of the Pythagorean triple to be determined is $v - u$; the two smaller terms are consecutive integers whose sum is $v + u$. Thus, 35 and 204 are adjacent terms whose difference is 169 and whose sum is $239 = 119 + 120$.

Exercise 3.5. Extend the definition of Pythagorean triple to involve negative as well as positive integers. $(-3, 4, 5)$ and $(8, 15, 17)$ are triples (a, b, c) for which $b = a + 7$. Show that the problem of finding other triples with this property leads to the equation $x^2 - 2y^2 = 7$. Determine solutions of this equation and derive from them more triples with the desired property.

Exercise 3.6. Show that there are infinitely many pairs $\{(a, b, c), (p, q, r)\}$ of primitive Pythagorean triples such that $|a - p|$, $|b - q|$, and $|c - r|$ are all equal to 3 or 4. (Problem 10704 in *American Mathematical Monthly* 106 (1999), 67; 107 (2000), 864).

Exploration 1.3. Just as $\sqrt{2}$ can be realized as the ratio of the hypotenuse of an isosceles right triangle to one of the equal sides, so $\sqrt{3}$ can be realized as the ratio of the longest side of an isosceles triangle with apex angle equal to 120° to one of the equal sides. In both cases, it is not possible to realize these as the ratios of the longest side to the equal side for similar triangles whose side lengths are integers. However, we can approximate the situation by an "almost isosceles" integer triangle with side lengths v, $v + 1$, and u, with the angle opposite the side u equal to 120°. We have the relation $3v^2 + 3v + 1 = u^2$, which can be written in the form $(2u)^2 - 3(2v + 1)^2 = 1$. Thus, we are led to solving the equation $x^2 - 3y^2 = 1$. Investigate solutions of this equation and determine the corresponding triangles. Does the ratio of u to v turn out to be a good approximation to $\sqrt{3}$?

Exploration 1.4. The quadratics $x^2+5x+6 = (x+2)(x+3)$ and $x^2+5x-6 = (x+6)(x-1)$ can both be factored as a product of linear polynomials with integer coefficients. Determine other integer pairs (m, n) for which $x^2 + mx + n$ and $x^2 + mx - n$ can both be so factored.

Exploration 1.5. This follows on from Exercise 3.6. For which positive integers a is it possible to find pairs of Pythagorean triples $[(x, y, z), (x + a, y + a, z + a + 1)]$?

1.4 Notes

See the monograph *The Great Mathematicians* by Herbert Westren Turnbull, reproduced in Volume I of *The World of Mathematics*, edited by James R. Newman, pages 75–168, for an account of the Greek contribution to the equation $x^2 - 2y^2 = 1$ (pages 97–98) in the context of sides and diagonals of squares. D.H. Fowler, in his book *The Mathematics of Plato's Academy: A New Reconstruction* (Oxford University Press, New York, 1987; second edition, 1999) discusses how ratio could be described and analyzed through the process of *anthyphairesis*, an iterative process akin to the Euclidean algorithm that produces for each magnitude a sequence of positive integers analogous to the sequence produced by the modern continued fraction algorithm. This process, applied to the diagonal and side of a square, is essentially the material given in Section 1.

2.3. The result of Exercise 2.3(a) is a special case of an important and general result on matrices, the Cayley–Hamilton theorem. We can define square matrices with n rows and n columns, where n is a positive integer, along with their addition and multiplication. Checking mechanically the analogue of the theorem, as we have done here, is not feasible, and in order to formulate and give an argument, we need to develop the theory of determinants and of characteristic polynomials, main topics in a first course on theoretical linear algebra.

1.5 Hints

1.1(a). Triangles ABF and AEF are right triangles with a common hypotenuse and a pair of corresponding equal sides.

1.7(d). Consider what the remainders of the terms are when divided by 8.

2.4(a). Use the fact that $p_n = p_{n-1} + 2q_{n-1}$ and $q_n = p_{n-1} + q_{n-1}$.

2.4(b). Observe that $p_{n+1}q_{n+1} = 6p_nq_n - p_{n-1}q_{n-1} + 2(p_{n-1}q_{n-1} + p_nq_{n-1} + p_{n-1}q_n - p_nq_n)$.

3.1(b). There are many ways of obtaining the general form for Pythagorean triples. One way is to note that each such triple consists of multiples of a triple a, b, c with greatest common divisor 1. Rewriting the Pythagorean equation as $a^2 = c^2 - b^2$, where c is odd and b is even, factor the right side; show that the factors are coprime and therefore perfect squares. An alternative argument that reveals more clearly the links with geometry and trigonometry is sketched below:

(i) Consider a right triangle ABC with $\angle C = 90^\circ$, arms a and b, and hypotenuse c. Let 2θ be the angle opposite the side b and $t = \tan\theta$. Verify that

$$CD = tBC,$$
$$CD : AD = CB : AB,$$
$$AD = tAB,$$
$$b = t(c+a),$$
$$t = \frac{b}{c+a} = \frac{c-a}{b},$$
$$\frac{a}{c} = \cos 2\theta = \frac{1-t^2}{1+t^2},$$
$$\frac{b}{c} = \sin 2\theta = \frac{2t}{1+t^2}.$$

(ii) Suppose that t is a rational number n/m, where m and n are positive integers. Verify that

$$\frac{a}{c} = \frac{m^2-n^2}{m^2+n^2},$$
$$\frac{b}{c} = \frac{2mn}{m^2+n^2},$$

and deduce that the given triangle is similar to one corresponding to the Pythagorean triple

$$(m^2-n^2, 2mn, m^2+n^2) \quad .$$

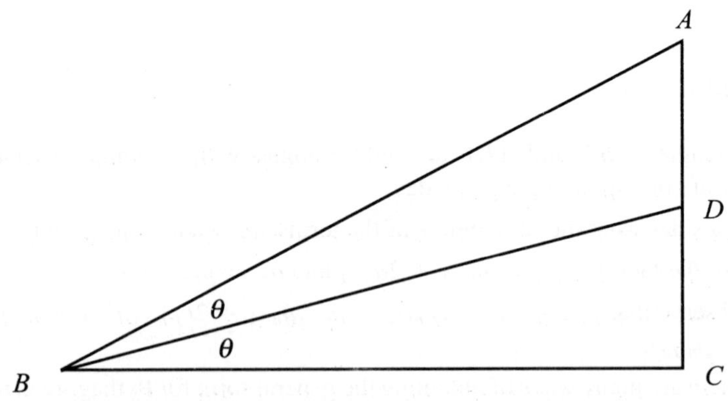

FIGURE 1.2.

(iii) Deduce from (a) and (b) that the triangles for which the quantity t is rational are precisely those that are similar to triangles corresponding to Pythagorean triples.

(iv) Suppose that (a, b, c) is a Pythagorean triple whose entries have greatest common divisor 1. Argue that exactly one of a and b is even; suppose it to be b. Let the corresponding value of t written as a fraction in lowest terms be n/m. Show that the greatest common divisor of $m^2 - n^2$, $2mn$, and $m^2 + n^2$ is either 1 or 2 and deduce that either

$$(a, b, c) = (m^2 - n^2, 2mn, m^2 + n^2)$$

or

$$(2a, 2b, 2c) = (m^2 - n^2, 2mn, m^2 + n^2).$$

The second possibility can be eliminated. Argue that when $m^2 \pm n^2$ are even, m and n must be odd, so that $m^2 - n^2$ is divisible by 4. Observe that this contradicts that a is odd.

(v) We have now demonstrated that every Pythagorean triple of numbers with greatest common divisor 1 has the form $(m^2 - n^2, 2mn, m^2 + n^2)$. Experiment with various values of m and n to obtain new Pythagorean triples. Which values of m and n will produce the triples you identified in Exercise 3.1(a)?

3.2(a). Complete the square of the quadratics $m^2 - n^2 - 2mn = m^2 - 2mn - n^2$ and $2mn - m^2 + n^2 = n^2 + 2mn - m^2$.

3.4. Consider $q_{n+1}^2 - q_n^2 - 2q_nq_{n+1}$. You will need to know something about $q_{n+1}q_{n-1} - q_n^2$.

3.6. Two examples are $\{(a, b, c), (p, q, r)\} = \{(5, 12, 13), (8, 15, 17)\}$, $\{(36, 77, 85), (39, 80, 89)\}$.

2

Problems Leading to Pell's Equation and Preliminary Investigations

The first chapter presented a situation that led to pairs of integers (x, y) that satisfied equations of the form $x^2 - 2y^2 = k$ for some constant k. One of the reasons for the popularity of Pell's equation as a topic for mathematical investigation is the fact that many natural questions that one might ask about integers lead to a quadratic equation in two variables, which in turn can be cast as a Pell's equation. In this chapter we will present a selection of such problems for you to sample.

For each of these you should set up the requisite equation and then try to find numerical solutions. Often, you should have little difficulty in determining at least one and may be able to find several. These exercises should help you gain some experience in handling Pell's equation. Before going on to study more systematic methods of solving them, spend a little bit of time trying to develop your own methods.

While a coherent theory for obtaining and describing the solutions of Pell's equation did not appear until the eighteenth century, the equation was tackled ingeniously by earlier mathematicians, in particular those of India. In the third section, inspired by their methods, we will try to solve Pell's equation.

2.1 Square and Triangular Numbers

The numbers 1, 3, 6, 10, 15, 21, 28, 36, 45, ... , $t_n \equiv \frac{1}{2}n(n+1)$, ... are called *triangular*, since the nth number counts the number of dots in an equilateral triangular array with n dots to the side.

It is not difficult to see that the sum of two adjacent triangular numbers is square.

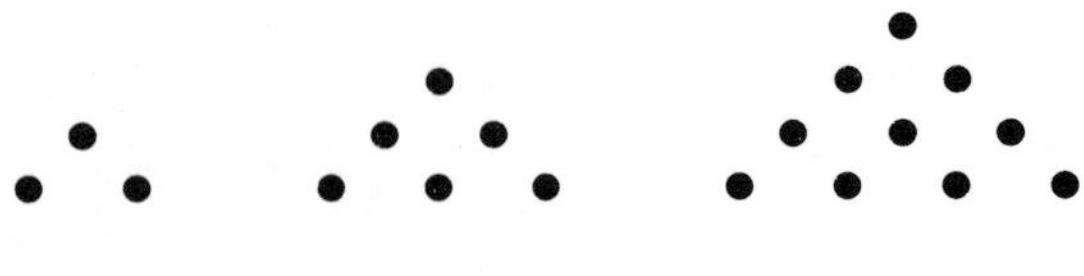

FIGURE 2.1.

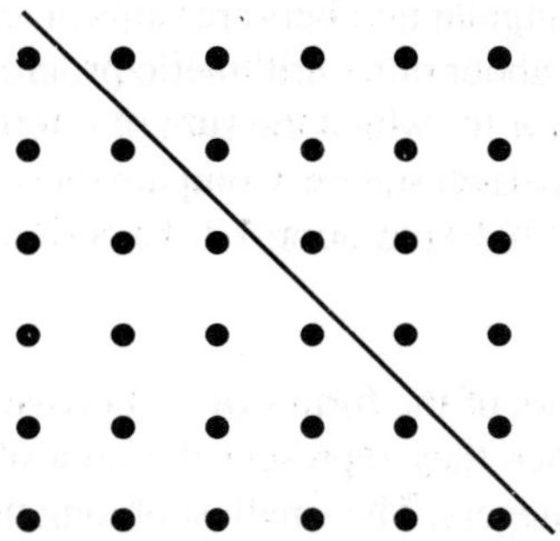

FIGURE 2.2.

But does it often happen that an individual triangular number is square? We will examine this and similar questions.

Exercise 1.1. Verify that the condition that the nth triangular number t_n is equal to the mth square is that $\frac{1}{2}n(n+1) = m^2$. Manipulate this equation into the form

$$(2n+1)^2 - 8m^2 = 1.$$

Thus, we are led to solving the equation $x^2 - 8y^2 = 1$ for integers x and y. It is clear that for any solution, x must be odd (why?), so that we can then find the appropriate values of m and n. Observe that 1 and 36 are included in the list of triangular numbers. What are the corresponding values of x, y, m, n? Use the results of Exercise 1.1.7(c) to generate other solutions.

Exercise 1.2. There are triangular numbers that differ from a square by 1, such as $3 = 2^2 - 1$, $10 = 3^2 + 1$, $15 = 4^2 - 1$, and $120 = 11^2 - 1$. Determine other examples.

Exercise 1.3. Find four sets of three consecutive triangular numbers whose product is a perfect square.

Exercise 1.4. Find four sets of three consecutive triangular numbers that add up to a perfect square.

Exercise 1.5. Determine integers n for which there exists an integer m for which $1 + 2 + 3 + \cdots + m = (m+1) + (m+2) + \cdots + n$.

Exercise 1.6. Determine positive integers m and n for which

$$m + (m+1) + \cdots + (n-1) + n = mn$$

(*International Mathematical Talent Search* 2/31).

Exploration 2.1. The triangular numbers are sums of arithmetic progressions. We can ask similar questions about other arithmetic progressions as well. Determine the smallest four values of n for which the sum of n terms of the arithmetic series $1+5+9+13+\cdots$ is a perfect square. Compare these values of n with the terms of the sequence $\{q_n\}$ listed in Exploration 1.1. Experiment with other initial terms and common differences.

Exploration 2.2. Numbers of the form $n(n+1)$ (twice the triangular numbers) are known as *oblong*, since they represent the area of a rectangle whose sides lengths are consecutive integers. The smallest oblong numbers are

$$2, 6, 12, 20, 30, 42, 56, 72, 90, 110, 132, 156.$$

A little experimentation confirms that the product of two consecutive oblong numbers is oblong; can you give a general proof of this result? Look for triples (a, b, c) of oblong numbers a, b, c for which $c = ab$. For each possible value of a, investigate which pairs (b, c) are possible.

An interesting phenomenon is the appearance of related triples of solutions. For example, we have (a, b, c) equal to

$$(14 \times 15, 782 \times 783, 11339 \times 11340),$$

$$(14 \times 15, 13 \times 14, 195 \times 196),$$

$$(13 \times 14, 782 \times 783, 10556 \times 10557),$$

while

$$(11339 \times 11340)(13 \times 14)^2 = (195 \times 196)(10556 \times 10557).$$

Are there other such triples?

2.2 Other Examples Leading to a Pell's Equation

The following exercises also involve Pell's equation. For integers n and k with $1 \leq k \leq n$, we define

$$\binom{n}{k} = \frac{n(n-1)\cdots(n-k+1)}{1 \cdot 2 \cdots k} = \frac{n!}{k!(n-k)!}.$$

Also, we define $\binom{n}{0} = 1$ for each positive integer n. Observe that $1+2+\cdots+n = \binom{n+1}{2}$.

Exercise 2.1. Determine nonnegative integers a and b for which

$$\binom{a}{b} = \binom{a-1}{b+1}.$$

Exercise 2.2. Suppose that there are n marbles in a jar with r of them are red and $n - r$ blue. Two marbles are drawn at random (without replacement). The probability that both have the same color is $\frac{1}{2}$. What are the possible values of n and r?

Exercise 2.3. The following problem appeared in the *American Mathematical Monthly* (#10238, 99 (1992), 674):

(a) Show that there exist infinitely many positive integers a such that both $a + 1$ and $3a + 1$ are perfect squares.
(b) Let $\{a_n\}$ be the increasing sequence of all solutions in (a). Show that $a_n a_{n+1} + 1$ is also a perfect square.

Exercise 2.4. Determine positive integers b for which the number $(111 \ldots 1)_b$ with k digits all equal to 1 when written to base b is a triangular number, regardless of the value of k.

Exercise 2.5. Problem 2185 in *Crux Mathematicorum* (22 (1996), 319) points out that

$$2^2 + 4^2 + 6^2 + 8^2 + 10^2 = 4 \cdot 5 + 5 \cdot 6 + 6 \cdot 7 + 7 \cdot 8 + 8 \cdot 9$$

and asks for other examples for which the sum of the first n even squares is the sum of n consecutive products of pairs of adjacent integers.

Exercise 2.6. Determine integer solutions of the system

$$2uv - xy = 16,$$
$$xv - uy = 12$$

(*American Mathematical Monthly* 61 (1954), 126; 62 (1955), 263).

Exercise 2.7. Problem 605 in the *College Mathematics Journal* (28 (1997), 232) asks for positive integer quadruples (x, y, z, w) satisfying $x^2 + y^2 + z^2 = w^2$ for which, in addition, $x = y$ and $z = x \pm 1$. Some examples are (2, 2, 1, 3) and (6, 6, 7, 11). Find others.

Exercise 2.8. The root-mean-square of a set $\{a_1, a_2, \ldots, a_k\}$ of positive integers is equal to

$$\sqrt{\frac{a_1^2 + a_2^2 + \cdots + a_k^2}{k}}.$$

Is the root-mean-square of the first n positive integers ever an integer? (USAMO, 1986)

Exercise 2.9. Observe that $(1 + 1^2)(1 + 2^2) = (1 + 3^2)$. Find other examples of positive integer triples (x, y, z) for which $(1 + x^2)(1 + y^2) = (1 + z^2)$.

Exercise 2.10. A problem in the *American Mathematical Monthly* (#6628, 98 (1991), 772–774) asks for infinitely many triangles with integer sides whose area is a perfect square. According to one solution, if m is chosen to make $\frac{1}{2}(m^2-1)$ a square, then the triangles with sides $(\frac{1}{2}(m^3+m^2)-1,\ \frac{1}{2}(m^3-m^2)+1,\ m^2)$ and with sides $(m^3-\frac{1}{2}(m-1),\ m^3-\frac{1}{2}(m+1),\ m)$ have square area. Recalling Heron's formula $\sqrt{s(s-a)(s-b)(s-c)}$ for the area of a triangle with sides (a, b, c) and perimeter $2s$, verify this assertion and give some numerical examples.

Exercise 2.11.

(a) Suppose that the side lengths of a triangle are consecutive integers $t-1, t, t+1$, and that its area is an integer. Prove that $3(t^2-4)$ must be an even perfect square, so that $t = 2x$ for some x. Thus show that $x^2-3y^2 = 1$ for some integer y. Determine some examples.

(b) In the situation of (a), prove that the altitude to the side of middle length is an integer and that this altitude partitions the side into two parts of integer length that differ by 4.

(c) Suppose that the sides of a triangle are integers $t-u$, t, and $t+u$. Verify that $3(t^2-4u^2)$ is a square $(3v)^2$ and obtain the equation $t^2-3v^2 = 4u^2$. Determine some examples with $u \neq 1$.

Exercise 2.12. Here is one approach to constructing triangles with integer sides whose area is an integer. Such a triangle can be had either by slicing one right triangle from another or by juxtaposing two right triangles. (See figure 2.3.)

We suppose that $m, r, a, b = ma + r$ and $c = ma - r$ are integers.

(a) Prove that $4mr = a \pm 2q$ and deduce that $2q$ is an integer.

(b) Prove that $2p$ must be an integer.

(c) By comparing two expressions for the area of the triangle (a, b, c), verify that

$$(4m^2-1)(a^2-4r^2) = 4p^2.$$

Take $a = 2t$ and obtain the equation

$$p^2-(4m^2-1)t^2 = -(4m^2-1)r^2.$$

(d) Determine some solutions of the equation in (c) and use them to construct some examples of triangles of the desired type.

Exercise 2.13. A Putnam problem (**A2** for the year 2000) asked for a proof that there are infinitely many sets of three consecutive positive integers each of which is the sum of two integer squares. An example of such a triple is $8 = 2^2+2^2$, $9 = 0^2+3^2$ and $10 = 1^2+3^2$.

(a) One way to approach the problem is to let the three integers be $n^2-1 = 2m^2$, n^2, and n^2+1. Derive a suitable Pell's equation for m and n and produce some numerical examples.

(b) However, it is possible to solve this problem without recourse to Pell's equation. Do this.

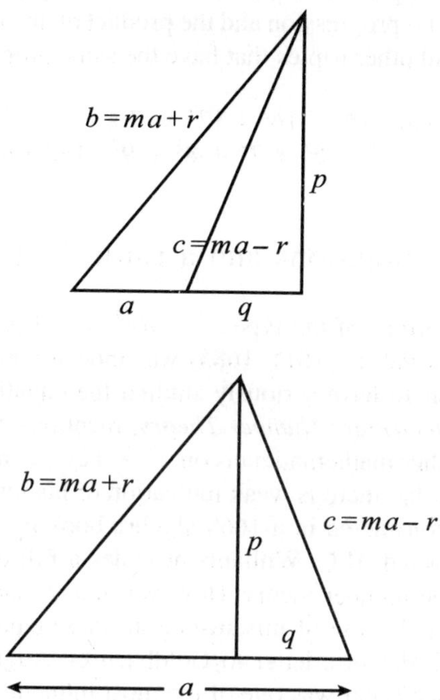

FIGURE 2.3.

Exploration 2.3. Let $\{F_n\}$ be the Fibonacci sequence determined by $F_0 = 0$, $F_1 = 1$, and $F_{n+1} = F_n + F_{n-1}$ for each integer n. It turns out that

$$F_{2n-1}^2 + F_{2n+1}^2 + 1 = 3F_{2n-1}F_{2n+1}$$

(can you prove this?), so that (F_{2n-1}, F_{2n+1}) is an example of a pair (a, b) for which $a^2 + b^2 + 1$ is a multiple of ab. Thus we have the instances $(a, b) = (1, 1), (1, 2), (2, 5), (5, 13), (13, 34)$. What other pairs can be found?

Problem #10316 in the *American Mathematical Monthly* (100 (1993), 589; 103 (1996), 905) asks for conditions under which ab divides $a^2 + b^2 + 1$. Suppose for some integer k that $a^2 + b^2 + 1 = kab$. If (a, b) satisfies the equation, then so also do $(b, kb - a)$ and $(a, ka - b)$, so we have a way of generating new solutions from old. Show that the multiple k must exceed 2, and that the Diophantine equation for a and b can be rewritten

$$(2a - kb)^2 - (k^2 - 4)b^2 = -4.$$

We can rule out certain values of k. For example, k must be a multiple of 3, but cannot be twice an odd number.

What are all the solutions for $k = 3$? Are there any solutions for $k = 15$? Are there solutions for any other values of k?

Exploration 2.4. The triple (1, 8, 15) has the interesting property that the three numbers are in arithmetic progression and the product of any two of them plus one is a perfect square. Find other triples that have the same property.

Exploration 2.5. Note that $15^2 + 16^2 + 17^2 = 7 \times (5^2 + 6^2 + 7^2)$ and $8^2 + 9^2 + 10^2 + 11^2 + 12^2 = 2 \times (5^2 + 6^2 + 7^2 + 8^2 + 9^2)$. Determine generalizations.

2.3 Strategies for Solutions and a Little History

It is perverse that equations of the type $x^2 - dy^2 = k$ became associated with the name of Pell. John Pell (c. 1611–1683) was indeed a minor mathematician, but he does not appear to have seriously studied the equation. Kenneth Rosen, on page 459 of his *Elementary Number Theory*, mentions a book in which Pell augmented work of other mathematicians on $x^2 - 12y^2 = n$, and D.E. Smith, in his *Source Book*, says that there is weak indication of his interest in the equation $x^2 = 12y^2 - 33$ is considered in a 1668 algebra book by J. H. Rahn to which Pell may have contributed. H.C. Williams provides a full description of this in his millennial paper on number theory. However, many mathematical historians agree that this is a simple case of misattribution; these equations were ascribed to Pell by Leonhard Euler in a letter to Goldbach on August 10, 1730, and in one of his papers. Since Euler was one of the most influential mathematicians in Europe in the eighteenth century, the name stuck. There were others very interested in the equation, many earlier than Pell. Pierre de Fermat (1601?–1665) was the first Western European mathematician to give the equation serious attention, and he induced his contemporaries John Wallis (1616–1703) and Frénicle de Bessy (1602–1675) to study it. Actually, Pell's equations go back a long way, before the seventeenth century. The Greeks seem to have come across some instances of it; in particular, Archimedes posed a problem about cattle that led to an equation of the type. The Indian mathematician Brahmagupta in the sixth century had a systematic way of generating infinitely many solutions from a particular one, while in the eleventh century, Jayadeva and Bhaskara II had algorithms for finding the first solution.

Exercise 3.1. In our definition of Pell's equation we specified that d had to be positive and nonsquare. Let us see why this restriction is a natural one. First, suppose that d is a negative number, say $-p$.

(a) Consider the equation $x^2 + 3y^2 = 7$. Find all solutions to this equation. How do you know that you have a complete set? In the Cartesian plane, sketch the curve with equation $x^2 + 3y^2 = 7$. Indicate all points on it with integer coordinates. What is this curve?

(b) For a given positive integer p and integer k, sketch the graph of the equation $x^2 + py^2 = k$. Corresponding to every point (x, y) on the graph with integer coordinates there is a solution to the equation. Determine an upper bound on the number of solutions that this equation can possibly have. ♠

Thus, when d is negative, the solutions of $x^2 - dy^2 = k$ in integers x and y are finite in number and can be found by inspection. Characterizing those values of k for which there *is* a solution for a given d is in itself an interesting question, though it is not within the scope of this book.

Exercise 3.2. Consider the equation $x^2 - dy^2 = k$ where $d = q^2$, the square of an integer.

(a) Determine all the solutions in integers that you can for each of the following equations:

$$x^2 - 4y^2 = 45,$$
$$x^2 - y^2 = 6,$$
$$x^2 - 9y^2 = 7.$$

(b) Argue that that equation $x^2 - q^2y^2 = k$ can have at most finitely many solutions in integers x and y. Give an upper bound for this number of solutions in terms of the number of positive integers that divide k evenly.

(c) Sketch the graph of the hyperbola with equation $x^2 - q^2y^2 = k$ along with the graphs of its asymptotes with equations $x + qy = 0$ and $x - qy = 0$. What are the points with integer coordinates lying on the asymptotes? What insight does this give as to why there are so few points on the hyperbola with integer coordinates?

Exercise 3.3. The eleventh century Indian mathematician Bhaskara was able to solve the equation $x^2 - 61y^2 = 1$ for integers x and y. One might think that since it easy to find a solution of $x^2 - 63y^2 = 1$ (do it!), there should not be too much difficulty solving Bhaskara's equation. However, simple trial and error is likely to lead to abject failure, and Bhaskara needed considerable numerical skill to handle the job—this at a time when there were no calculators or even the convenient notation we enjoy today. In this exercise we will indicate the type of strategy followed by Bhaskara, but avail ourselves of modern notation.

(a) Suppose that k, d, x, y are integers for which $x^2 - dy^2 = k$. Show that

$$(mx + dy)^2 - d(ym + x)^2 = k(m^2 - d)$$

for each positive integer m.

(b) Suppose, in (a), that the greatest common divisor of k and y is 1, and that $ym + x$ is a multiple of k. Use the equations

$$(m^2 - d)y^2 = k - (x^2 - m^2y^2) = k - (x + my)(x - my)$$

and

$$mx + dy = m(x + my) - (m^2 - d)y$$

to show that $m^2 - d$ and $mx + dy$ are also multiples of k. Thus, the result in (a) can be rewritten

$$\left(\frac{mx + dy}{k}\right)^2 - d\left(\frac{ym + x}{k}\right)^2 = \frac{m^2 - d}{k},$$

where the quantities in parentheses can be made all integers when m is suitably chosen.

(c) Derive from $8^2 - 61(1)^2 = 3$ the equation

$$\left(\frac{8m + 61}{3}\right)^2 - 61\left(\frac{m + 8}{3}\right)^2 = \frac{m^2 - 61}{3}.$$

Choose m so that $(m + 8)/3$ is an integer and $|(m^2 - 61)/3|$ is as small as possible. Hence derive

$$39^2 - 61(5)^2 = -4.$$

(d) Now obtain the equation

$$\left(\frac{39m + 305}{-4}\right)^2 - 61\left(\frac{5m + 39}{-4}\right)^2 = \frac{m^2 - 61}{-4}.$$

Choose m so that $(5m + 39)/4$ is an integer and $|(m^2 - 61)/4|$ is as small as possible. Using a pocket calculator, if you wish, obtain

$$164^2 - 61(21)^2 = -5.$$

(e) We can continue on in this way to successively derive the following numerical equations. Check the derivation of as many of them as you need to feel comfortable with the process.

$$\begin{aligned}
453^2 - 61(58)^2 &= 5,\\
1523^2 - 61(195)^2 &= 4,\\
5639^2 - 61(722)^2 &= -3,\\
29718^2 - 61(3805)^2 &= -1,\\
469849^2 - 61(60158)^2 &= -3,\\
2319527^2 - 61(296985)^2 &= 4,\\
9747957^2 - 61(1248098)^2 &= 5,\\
26924344^2 - 61(3447309)^2 &= -5,\\
90520989^2 - 61(11590025)^2 &= -4,\\
335159612^2 - 61(42912791)^2 &= 3,\\
1766319049^2 - 61(226153980)^2 &= 1.
\end{aligned}$$

It is interesting to note that the equation $x^2 - 61y^2 = 1$ was proposed by the Frenchman Pierre Fermat to Frénicle in February, 1657. The first European to publish a solution was Leonhard Euler, in 1732.

Exercise 3.4. There are devices known to Bhaskara by which the process can be shortened. One depends on the identity

$$(x^2 - dy^2)(u^2 - dv^2) = (xu + dyv)^2 - d(xv + yu)^2.$$

(a) Verify this identity and draw from it the conclusion that if two integers can be written in the form $x^2 - dy^2$ for integers x and y, then so can their product.
(b) Explain how a solution of $x^2 - dy^2 = -1$ can be used to obtain a solution of $x^2 - dy^2 = +1$.
(c) Determine a solution to the equation $x^2 - 65y^2 = 1$ in integers x and y.
(d) From the identity and the pair of equations, derive

$$x^2 - dy^2 = k, \qquad m^2 - d(1)^2 = m^2 - d,$$

the equation in Exercise 3.3(a).

Exercise 3.5. Refer to Exercise 3.3.
(a) From the numerical equation $39^2 - 61(5)^2 = -4$, deduce that

$$\left(\frac{39}{2}\right)^2 - 61\left(\frac{5}{2}\right)^2 = -1.$$

(b) Substituting $x = u = 39/2$, $y = v = 5/2$, $d = 61$ in the identity of Exercise 3.4(a), derive

$$\left(\frac{1523}{2}\right)^2 - 61\left(\frac{195}{2}\right)^2 = 1.$$

(c) Substituting $x = 39/2$, $y = 5/2$, $u = 1523/2$, $v = 195/2$ in the identity of Exercise 3.4(a), obtain

$$29718^2 - 61(3805)^2 = -1.$$

(d) Now obtain a solution to $x^2 - 61y^2 = 1$ using Exercise 3.4(b).

Exercise 3.6. Another equation solved by Bhaskara was $x^2 - 67y^2 = 1$.
(a) Following the procedure of Exercise 3.3, derive the equations

$$\begin{aligned} 8^2 - 67(1)^2 &= -3, \\ 41^2 - 67(5)^2 &= 6, \\ 90^2 - 67(11)^2 &= -7, \\ 221^2 - 67(27)^2 &= -2. \end{aligned}$$

(b) Using the identity of Exercise 3.4, derive a solution of $x^2 - 67y^2 = 4$ and deduce from this a solution in integers x, y to $x^2 - 67y^2 = 1$.

Exercise 3.7. In 1658, Frénicle claimed that he had found a solution in integers x and y to $x^2 - dy^2 = 1$ for all nonsquare values of d up to 150, but mentioned that he was looking in particular for solutions in the cases $d = 151$ and $d = 313$. In response, John Wallis found that

$$(1728148040)^2 - 151(140634693)^2 = 1,$$

and Lord Brouncker commented that within an hour or two, he had discovered that

$$(126862368)^2 - 313(7170685)^2 = -1.$$

Check that these results are correct. Doing this in this obvious way may not be most efficient, particularly if they lead to overflow of your pocket calculator. A better way may be to set things up so that you can use division rather than multiplication. This might involve manipulating the equation to be checked into forms leading to easy factorization, such as those involving differences of squares. It might involve checking for small prime factors of terms involved. Be creative and use some ingenuity.

Exercise 3.8. How might Wallis and Brouncker have solved a Pell's equation? Consider the example $x^2 - 7y^2 = 1$.

(a) The smallest square exceeding 7 is $9 = 3^2$; we have that $7 = 3^2 - 2$. Deduce from this $7(2)^2 = 6^2 - 8$, $7(3)^2 = 9^2 - 18$, and more generally $7m^2 = (3m)^2 - 2m^2$.

(b) Observe that $9^2 - 18 = (9-1)^2 - 1$, so that $9^2 - 18$ is just 1 shy of being a perfect square. Transform $7(3)^2 = 9^2 - 18$ to $7(3)^2 = (9-1)^2 - 1$ and thence derive a solution to $x^2 - 7y^2 = 1$.

Exercise 3.9. Let us apply the Wallis–Brouncker approach to the general equation $x^2 - dy^2 = 1$. Let positive integers c and k be chosen to satisfy $(c-1)^2 < d < c^2$ and $c^2 - d = k$, so that c^2 is the smallest square exceeding d, and k is the difference between this square and d.

(a) Show that for any integer m, $dm^2 = (cm)^2 - km^2$.

(b) The quantity $dm^2 = (cm)^2 - km^2$ is certainly less than $(cm)^2$, but it may not be less than any smaller square, in particular $(cm-1)^2$. However, as m grows larger, the distance between dm^2 and $(cm)^2$ increases, so that eventually dm^2 will become less than $(cm-1)^2$. This will happen as soon as

$$(cm)^2 - km^2 \leq (cm-1)^2 - 1.$$

Verify that this condition is equivalent to $2c \leq km$.

(The strategy is to select the smallest value of m for which this occurs and hope that $dm^2 - (cm-1)^2 = -1$, in which case $(x, y) = (cm-1, m)$ will satisfy $x^2 - dy^2 = 1$. This, of course, need not occur, and we will need to modify the strategy.)

(c) Start with $13 = 4^2 - 3$ and determine the smallest value of m for which $13m^2 - (4m-1)^2$ has a negative value, and write the numerical equation that evaluates this.

(d) Suppose $dm^2 - (cm-1)^2$ is not equal to -1 when it first becomes negative. It will take larger and larger negative values as m increases (why?). Eventually, there will come a time when $dm^2 - (cm - 2)^2$ will not exceed -1. Verify that this occurs when $4c \le km + (3/m)$.

(e) This process can be continued. If $dm^2 - (cm - 2)^2$ fails at any point to equal -1, then we can try for a solution with $x = cm - 3$ for some value of m. This process can be continued until (hopefully) a solution is found. Try this in the case that $d = 13$.

(f) Which method do you consider more convenient, this or Bhaskara's?

Exercise 3.10. Here is a systematic way to obtain solutions of $x^2 - dy^2 = 1$ for a great many values of d.

(a) Verify the identity

$$(zy + 1)^2 - \left(z^2 + \frac{2z}{y}\right)y^2 = 1.$$

(b) Suppose that integers y and z are selected so that $2z$ is a multiple of y; let $d = z^2 + (2z/y)$ and $x = zy + 1$. Without loss of generality, we may suppose that $z > 0$ and that y can be either positive or negative. If $1 \le y \le 2z$, show that $z^2 + 1 \le d \le (z + 1)^2 - 1$, while if $-2z \le y \le -1$, show that $(z - 1)^2 - 1 \le d \le z^2 + 1$.

(c) Describe how, for a given value of d, one *might* determine solutions to $x^2 - dy^2 = 1$. Apply this method to obtain solutions when $d = 3, 27, 35, 45$.

(d) List values of d up to 50 for which solutions cannot be found using this method.

2.4 Explorations

Exploration 2.6. Archimedes' Cattle Problem. In the eighteenth century, a German dramatist, G.E. Lessing, discovered a problem posed by Archimedes to students in Alexandria. A complete statement of the problem and comments on its history and solution can be found in the following sources:

H.W. Lenstra, Jr., Solving the Pell's equation. *Notices of the American Mathematical Society* 49:2 (February, 2002), 182-192.

James R. Newman (*ed.*), *The World of Mathematics, Volume 1* (Simon & Schuster, New York, 1956) pages 197–198, 105–106.

H.L. Nelson, A solution to Archimedes' cattle problem, *Journal of Recreational Mathematics* 13:3 (1980–81), 162–176.

Ilan Vardi, Archimedes' Cattle Problem, *American Mathematical Monthly* 106 (1998), 305–319.

The paper of H.C. Williams on solving Pell's equation, delivered to the Millennial Conference on Number Theory in 2002 and listed in the historical references, discusses the cattle problem and lists additional references by P. Schreiber and W.

Waterhouse. In modern symbolism, this problem amounts to finding eight positive integers to satisfy the conditions

$$W = \tfrac{5}{6}X + Z, \quad X = \tfrac{9}{20}Y + Z, \quad Y = \tfrac{13}{42}W + Z,$$
$$w = \tfrac{7}{12}(X + x), \quad x = \tfrac{9}{20}(Y + y), \quad y = \tfrac{11}{30}(Z + z), \quad z = \tfrac{13}{42}(W + w),$$

with the additional requirements that $W + X$ is to be square and $Y + Z$ triangular. Solving this problem involves obtaining a solution to the Pell's equation $p^2 - 4{,}729{,}494q^2 = 1$, a feat that was not accomplished until 1965. Now, of course, we have sophisticated software available to do the job. Can you find a solution to the equation?

Exploration 2.7. There are certain values of d for that it is easy to find a solution of $x^2 - dy^2 = 1$. One does not have to look very far to solve $x^2 - 3y^2 = 1$ or $x^2 - 8y^2 = 1$. Indeed, there are categories of values of d for which some formula for a solution can be given. For example, 3 and 8 are both of the form $t^2 - 1$; what would a solution of $x^2 - (t^2 - 1)y^2 = 1$ be for an arbitrary value of the parameter t? Can you find more than one solution?

Determine the smallest pair (x, y) of positive integers that satisfies $x^2 - dy^2 = 1$ in each of the following special cases.

(a) $d = 2, 5, 10, 17, 26, \ldots, t^2 + 1, \ldots$.
(b) $d = 3, 6, 11, 18, 27, \ldots, t^2 + 2, \ldots$.
(c) $d = 2, 7, 14, 23, \ldots, t^2 - 2, \ldots$.
(d) $d = 2, 6, 12, 20, 30, \ldots, t^2 + t, \ldots$.
(e) $d = 7, 32, 75, \ldots, t^2 + (4t + 1)/3, \ldots$ (where t is 1 less than a multiple of 3.
(f) $d = 3, 14, 33, \ldots, t^2 + (3t + 1)/2, \ldots$ (where t is odd).

Now we come to some tougher cases that do not seem to follow an easy pattern. Find at least one solution in positive integers to each of the following:

(g) $x^2 - 21y^2 = 1$.
(h) $x^2 - 22y^2 = 1$.
(i) $x^2 - 28y^2 = 1$.
(j) $x^2 - 19y^2 = 1$.
(k) $x^2 - 13y^2 = 1$.
(l) $x^2 - 29y^2 = 1$.
(m) $x^2 - 31y^2 = 1$.

Exploration 2.8. For which values of the integer d is $x^2 - dy^2 = -1$ solvable? In particular, is there a solution when d is a prime exceeding a multiple of 4 by 1? Do not look at the discussion for this exploration until you have completed working through Chapter 5.

Exploration 2.9. Which integers can be written in the form $x^2 - y^2$? $x^2 - 2y^2$? $x^2 - 3y^2$? $x^2 - dy^2$?

Exploration 2.10. Let $x_n = a + (n-1)d$ be the nth term of an arithmetic progression with initial term a and common difference d. The quantity $s_n = x_1 + x_2 + \cdots + x_n$ is called a partial sum of the series $x_1 + x_2 + x_3 + \cdots$. Must there be at least one partial sum that is a square? Even if the a and d are coprime? For which progressions is it true that every partial sum is a square? Suppose that there is one square partial sum; must there be infinitely many more?

Exploration 2.11. In Exploration 1.4 the equation $x^2 - 3y^2 = 1$ was considered. Its solutions are given by

$$(x, y) = (1, 0), (2, 1), (7, 4), (26, 15), (97, 56), (362, 209), \ldots.$$

What is special about the numbers 2, 26, and 362? The solutions for $x^2 - 6y^2 = 1$ are

$$(x, y) = (1, 0), (5, 2), (49, 20), (485, 198), (4801, 1960), \ldots.$$

Note the appearance of 5 and 485. You may also wish in this context to look at the solutions of $x^2 - 7y^2 = 1$ and $x^2 - 8y^2 = 1$. Are there other values of d for which the solutions of $x^2 - dy^2 = 1$ exhibit similar behavior?

2.5 Historical References

There are several books and papers concerned with the history of Pell's equation:

David M. Burton, *The History of Mathematics: An Introduction*. Allyn and Bacon, Newton, MA, 1985 [pp. 243, 250, 504].

Bibhutibhusan Datta and Avadhesh Narayan Singh, *History of Hindu Mathematics, A Source Book*, Asia Publishing House, Bombay, 1962.

Leonard Eugene Dickson, *History of the Theory of Numbers, Volume II: Diophantine Analysis*. Chelsea, New York, 1952 (reprint of 1920 edition) [Chapter XII].

Victor J. Katz, *A History of Mathematics: An Introduction*. (Harper-Collins, New York, 1993) [pp. 208–211, 555–556].

Morris Kline, *Mathematical Thought from Ancient to Modern Times*. Oxford University Press, New York, 1972 [pp. 278, 610, 611].

James R. Newman *editor, The World of Mathematics, Volume 1*. Simon and Schuster, New York, 1956 [pp. 197–198].

C.O. Selenius, Rationale of the Chakravala process of Jayadeva and Bhaskara II, *Historia Mathematica* 2 (1975), 167–184.

David E. Smith *ed., A Source Book in Mathematics, Volume One* Dover, 1959 [pp. 214–216].

D.J. Struik *ed.*, *A Source Book in Mathematics, 1200–1800*, Harvard University Press, Cambridge, MA, 1969 [pages 29–31].

André Weil, *Number Theory: An Approach Through History from Hammurapi to Legendre*. Birkhäuser, Boston, 1983.

H.C. Williams, Solving the Pell's equation. *Proceedings of the Millennial Conference on Number Theory* (Urbana, IL, 2000) (M.A. Bennett *et al.*, editors), A.K. Peters, Boston, 2002.

Weil describes how rational approximations to the square root of 3 involved obtaining some solutions to Pell's equation. Burton, Dickson (pp. 342–345), and Newman mention the Archimedean cattle problem. Datta and Singh, Dickson (pp. 346–350), Katz, and Weil give quite a bit of attention to Indian mathematics, with Selenius giving an analysis of their method. Dickson and Weil give a lot of detail on European developments in the seventeenth and eighteenth centuries. Smith and Struik document a 1657 letter of Fermat in which he asserts that given any number not a square, there are infinitely many squares that when multiplied by the given number are one less than a square.

2.10, 2.11. See A.R. Beauregard and E.R. Suryanarayan, Arithmetic triangle, *Mathematics Magazine* (1997) 106–116.

5.9. *Oeuvres de Fermat* III, 457-480, 490-503; Dickson, p. 352.

2.6 Hints

2.3(a). If the two numbers are y^2 and x^2, what is $x^2 - 3y^2$?

2.4. In particular, $1 + b + b^2 = \frac{1}{2}v(v+1)$ for some integer v.

2.5. Suppose that $2^2 + 4^2 + \cdots + (2n)^2 = m(m+1) + \cdots + (m+n-1)(m+n)$. The left side can be summed using the formula for the sum of the first r squares: $\frac{1}{6}r(r+1)(2r+1)$. The right side can be summed by expressing each term as a difference: $3x(x+1) = x(x+1)(x+2) - (x-1)x(x+1)$. Show that this equation leads to $(n+1)^2 = m(n+m)$, which can be rewritten as a Pythagorean equation: $n^2 + [2(n+1)]^2 = (2m+n)^2$. At this point you can use the general formula for Pythagorean triples to get an equation of the form $x^2 - 5y^2 = 4$.

2.6. Square each equation and eliminate terms that are linear in each variable.

3.7. Rearrange the terms in the equation involving 151 to obtain a difference of squares on one side. The equation involving 313 is trickier. It may help to observe that $313 = 12^2 + 13^2$. It is easy to check divisibility of factors by powers of 2. Casting out 9's will help check divisibility by powers of 3. Check also for divisibility of factors by other small primes. Another way to compare divisors is as follows. Suppose we wish to show that $ab = cd$. We might look for common divisors of the two sides; one such would be the greatest common divisor of a and c. Finding such a greatest common divisor need not involve knowing the prime-power decomposition of the numbers. The Euclidean algorithm can be used.

Suppose $a > c$. Divide c into a and get a remainder $r = a - cq$, where q is the quotient. Then $\gcd(a, c) = \gcd(c, r)$. Now we have a smaller pair of numbers to work with. We can continue the process with c and r. Eventually, we will come to a pair of numbers one of which divides the other.

3

Quadratic Surds

Once a solution in integers to Pell's equation $x^2 - dy^2 = 1$ is given, it is possible to generate infinitely many others. Underlying this is an algebraic structure that can be revealed through operations on surds.

3.1 Quadratic Surds

Let a, b, d be rational numbers with d a nonsquare positive integer. A *quadratic surd* is a number of the form $a + b\sqrt{d}$; its *surd conjugate* is $\overline{a + b\sqrt{d}} = a - b\sqrt{d}$. Multiplying the quadratic surd by its conjugate gives its *norm* $N(a + b\sqrt{d}) = a^2 - b^2d$. Since the norm has the same form as the left side of Pell's equation, it is not surprising that surds have a role to play in the analysis of that equation.

Exercise 1.1.

(a) Write the product of $2 + 7\sqrt{3}$ and $3 - 4\sqrt{3}$ in the form $a + b\sqrt{3}$, where a and b are integers.

(b) What are the norms of $2 + 7\sqrt{3}$, $3 - 4\sqrt{3}$, and the product of these surds?

(c) Write $(2 + 7\sqrt{3})^{-1}$ in the form $u + v\sqrt{3}$, where u and v are rational numbers.

(d) Observe that $2 + \sqrt{3}$ has a multiplicative inverse $p + q\sqrt{3}$, where p and q are not merely rationals, but integers. Determine other numbers of the form $a + b\sqrt{3}$, where a and b are integers, whose multiplicative inverses also have integer coefficients.

Exercise 1.2. Let $c = a + b\sqrt{d}$ and $w = u + v\sqrt{d}$. Verify that:

(a) $\overline{cw} = \bar{c} \times \bar{w}$.

(b) $N(cw) = N(c)N(w)$.

(c) $N(c + w) + N(c - w) = 2(N(c) + N(w))$.

(d) $c/w = (c\bar{w})/(N(w))$.

Exercise 1.3. Verify that Pell's equation $x^2 - dy^2 = k$ can be written in the form $N(x + y\sqrt{d}) = k$.

Exercise 1.4. Suppose that $x^2 - dy^2 = k$ and $u^2 - dv^2 = l$ are two given integer equations. Define the integers m and n by

$$m + n\sqrt{d} = (x + y\sqrt{d})(u + v\sqrt{d}).$$

Verify that $m = xu + dyv$, $n = xv + yu$, and $m^2 - dn^2 = kl$.

Exercise 1.5.

(a) Suppose that $(x, y) = (x_1, y_1)$ is a solution to $x^2 - dy^2 = 1$. Define the integer pair (x_2, y_2) by the equation

$$x_2 + y_2\sqrt{d} = (x_1 + y_1\sqrt{d})^2.$$

Verify that $x_2 = x_1^2 + dy_1^2$, $y_2 = 2x_1y_1$ and (x_2, y_2) is a solution of $x^2 - dy^2 = 1$.

(b) More generally, suppose that (x_n, y_n) is defined by

$$(x_n + y_n\sqrt{d}) = (x_1 + y_1\sqrt{d})^n \quad \text{for } n \geq 2.$$

Note that $x_n + y_n\sqrt{d} = (x_{n-1} + y_{n-1}\sqrt{d})(x_1 + y_1\sqrt{d})$ and deduce that

$$x_n = x_1x_{n-1} + dy_1y_{n-1}, \qquad y_n = x_1y_{n-1} + y_1x_{n-1}.$$

Prove that (x_n, y_n) is a solution of $x^2 - dy^2 = 1$.

(c) Deduce that if $x^2 - dy^2 = 1$ has a solution other than $(x, y) = (\pm 1, 0)$, then it has infinitely many solutions in positive integers.

Exercise 1.6. Given that $(x, y) = (3, 2)$ is a solution of $x^2 - 2y^2 = 1$, generate a sequence of solutions by applying Exercise 1.5.

Exercise 1.7. Given that $(x, y) = (1, 1)$ is a solution of $x^2 - 2y^2 = -1$, generate an infinite sequence of solutions to this equation.

Exercise 1.8. Given that $(x, y) = (3, 1)$ is a solution of $x^2 - 2y^2 = 7$, generate an infinite sequence of solutions to this equation.

Exercise 1.9. The surd technique can be used in a more general way. Consider problem #2219 from *Crux Mathematicorum*: *Show that there are an infinite number of solutions in integers of the simultaneous equations*

$$x^2 - 1 = (u + 1)(v - 1),$$
$$y^2 - 1 = (u - 1)(v + 1).$$

(a) Suppose that the two equations are satisfied. Deduce that

$$(v + 1)x^2 - (v - 1)y^2 = 2v^2.$$

(b) Making the substitution $v = 2w$, $r = x/v$, $s = y/v$, obtain from (a) that

$$\left(w + \frac{1}{2}\right)r^2 - \left(w - \frac{1}{2}\right)s^2 = 1.$$

(c) Let w be arbitrary. Determine a simple numerical solution of the equation in (b).

(d) Observe that the odd positive integer powers of $r\sqrt{w+\frac{1}{2}}+s\sqrt{w+\frac{1}{2}}$ have the form $R\sqrt{w+\frac{1}{2}}+S\sqrt{w+\frac{1}{2}}$.

(e) Explain how (c) and (d) lead to the result that if $(r,s)=(r_0,s_0)$ satisfies the equation in (b), then so also does $(r,s)=(r_n,s_n)$ defined recursively by

$$r_{n+1}=2wr_n+(2w-1)s_n,$$
$$s_{n+1}=(2w+1)r_n+2ws_n.$$

(f) Using (e) and the transformation $x_n=r_nv$ and $y_n=s_nv$, prove that if (x_0,y_0) satisfies the equation in (a), then so also does $(x,y)=(x_n,y_n)$ defined recursively for $n\geq 0$ by

$$x_{n+1}=vx_n+(v-1)y_n,$$
$$y_{n+1}=(v+1)x_n+vy_n.$$

(g) Verify that if $(v+1)x_n^2-(v-1)y_n^2=2v^2$ and $|v|\neq 1$, then

$$\frac{x_n^2-v}{v-1}=\frac{y_n^2+v}{v+1}.$$

(h) Verify that

$$y_{n+1}^2+v=(v+1)[(v+1)x_n^2+2vx_ny_n+vy_n^2+v]-v(y_n^2+v).$$

(i) Let v be a fixed integer, and suppose that $(x,y)=(x_0,y_0)$ is a solution of the equation in (a) for which y_0^2+v is a multiple of $v+1$. With (x_n,y_n) as already defined, let

$$u_n=\frac{y_n^2+v}{v+1}.$$

Prove that $(x,y,u,v)=(x_n,y_n,u_n,v)$ is a quadruple of integers satisfying the system given in the problem.

(j) For each integer value of v, explicitly display an infinite system of solutions of the system. What happens if $v=1$?

Exercise 1.10.

(a) Let m be a positive integer. Determine a positive integer r such that $mr+1$ and $(m+1)r+1$ are both perfect squares.

(b) Suppose that $mr+1=x^2$ and $(m+1)r+1=y^2$. Verify that $(m+1)x^2-my^2=1$. On the other hand, if $(m+1)x^2-my^2=1$, deduce that m must divide x^2-1 and show how a solution of the system

$$mr+1=x^2 \quad\text{and}\quad (m+1)r+1=y^2$$

can be found.

(c) Determine a solution of $(m+1)x^2-my^2=1$.

(d) One way to find an infinite sequence of solutions to the equation in (c) is to consider odd powers of

$$\sqrt{m+1} + \sqrt{m}.$$

(Why?) Suppose that this generates sequences $\{x_n\}$, $\{y_n\}$, and $\{r_n\}$ for which

$$mr_n + 1 = x_n^2 \quad \text{and} \quad (m+1)r_n + 1 = y_n^2.$$

Write out the first few terms of this sequence.

Exploration 3.1. In Exercise 1.10, try to find a recursion relation for the sequence $\{r_n\}$.

3.2 Existence of Rational Solutions

It is natural to ask whether $x^2 - dy^2 = 1$ has a solution other than the trivial $(x, y) = (\pm 1, 0)$ when d is a positive nonsquare integer. The empirical investigation in Chapter 2 suggests an affirmative answer, although for some values of d the search was arduous. Here is a table of smallest solutions for low values of d:

d	(x, y)	d	(x, y)
2	(3, 2), (17, 12)	10	(19, 6)
3	(2, 1), (7, 4)	11	(10, 3)
5	(9, 4), (161, 72)	12	(7, 2), (97, 28)
6	(5, 2), (49, 20)	13	(649, 180)
7	(8, 3), (127, 48)	14	(15, 4)
8	(3, 1), (17, 6)	15	(4, 1), (31, 8)

One way to approach the situation is to ask first for solutions that are rational, and perhaps use these to obtain integer solutions. In this case, the task becomes much easier.

Exercise 2.1. We first establish that $x^2 - dy^2$ can assume square values for suitable integers x and y. Observe that if c is any integer, then when $(x, y) = (c, 1)$, then $x^2 - dy^2$ assumes the value $k = c^2 - d$.

(a) Determine a solution of the equation $x^2 - dy^2 = (c^2 - d)^2$.

(b) Describe how to determine solutions of $x^2 - dy^2 = 1$ for which x and y are rational, but not necessarily integers.

Exercise 2.2.

(a) Solve in positive integers the equation $x^2 - 13y^2 = 3^2$ and derive a rational solution of $x^2 - 13y^2 = 1$.

(b) Find the smallest solution in positive integers of $x^2 - 13y^2 = 2^2$ and derive the rational solution $(x, y) = (\frac{11}{2}, \frac{3}{2})$ of $x^2 - 13y^2 = 1$.

(c) By considering $(11 + 3\sqrt{13})^2$, determine a second rational solution $(x, y) = (\frac{119}{2}, \frac{33}{2})$ of $x^2 - 13y^2 = 1$.

(d) From (b) and (c), we have $11^2 - 13 \times 3^2 = 4$ and $119^2 - 13 \times 33^2 = 4$. Write

$$\frac{119 + 33\sqrt{13}}{11 - 3\sqrt{13}}$$

in the form $u + v\sqrt{13}$ where u and v are rational. Verify that in fact, u and v are integers for which $u^2 - 13v^2 = 1$.

Exercise 2.3. Here is another way to solve $x^2 - 13y^2 = -1$.

(a) Observe that $4^2 - 13 \times 1^2 = 3$, while $7^2 - 13 \times 2^2 = -3$. From these facts, determine a solution of $x^2 - 13y^2 = -9$.

(b) From (a), determine a solution in integers for $x^2 - 13y^2 = -1$.

Exercise 2.4.

(a) From the equation $8^2 - 61 \times 1^2 = 3$ and $7^2 - 61 \times 1^2 = -2^2 \times 3$, determine a solution in integers for $x^2 - 61y^2 = -4$.

(b) Solve $x^2 - 61y^2 = -1$ in rational numbers and use this to determine a solution in integers.

Exercise 2.5. Let d be a nonsquare positive integer. Consider the set of numbers

$$D = \{x + y\sqrt{d} : x \text{ and } y \text{ are rational and } x^2 - dy^2 = 1\}.$$

Verify that for this set of numbers, the product and quotient of any pair of nonzero numbers in the set also belong to the set. This is often described as closure under multiplication and division.

Exercise 2.6.

(a) Taking $c = 5$ and $d = 19$ in Exercise 2.1, obtain the solution $(x, y) = (22, 5)$ for $x^2 - 19y^2 = 9$.

(b) Taking $c = 4$ and $d = 19$, obtain the solution $(x, y) = (35, 8)$ for $x^2 - 19y^2 = 9$.

(c) By considering the quotients of $22 \pm 5\sqrt{19}$ and $35 \pm 8\sqrt{19}$, obtain a solution in integers (x, y) for $x^2 - 19y^2 = 1$.

Exercise 2.7. Follow the procedures of the previous exercise to determine solutions for $x^2 - dy^2 = 1$ when $d = 21, 22, 28$.

Exercise 2.8. We can follow the strategy of Exercise 2.3 in finding rational solutions for $x^2 - dy^2 = 1$. Letting $x = u/w$ and $y = v/w$, we obtain $u^2 - w^2 = dv^2$.

(a) Consider the case $d = 2$. By factoring the left side, argue that v must be even and that if we can take v to be an arbitrary product $2rs$, then we can set $u - w = 2r^2, u + w = 4s^2$ or $u - w = 4r^2, u + w = 2s^2$. Obtain the parametric solutions

$$(u, v, w) = (r^2 + 2s^2, 2rs, 2s^2 - r^2)$$

and

$$(u, v, w) = (2r^2 + s^2, 2rs, s^2 - 2r^2).$$

(b) Obtain a parametric set of solutions for $u^2 - dv^2 = w^2$.

Exercise 2.9. Here is a recent method due to Charles Galloway, a Toronto actuary, for obtaining rational and possibly integer solutions for $x^2 - dy^2 = 1$. Write $d = k^2 + c$, where k^2 is not necessarily the nearest square to d nor even an integer.

(a) Verify that $(x, y) = (1 + 2k^2/c, 2k/c)$ is a solution.

(b) We try to find values of k such that $2k/c = 2k/(d - k^2)$ is an integer. For $d = 13$, verify that $k = 3.6$ works and yields integer values of x and y. (Cf. Exercise 2.3.10.)

Exploration 3.2. Try the Galloway method of Exercise 2.9 to obtain solutions of $x^2 - dy^2 = 1$ for other values of d.

3.3 "Powers" of a Solution

Suppose d is a nonsquare positive integer and that $(x, y) = (u, v)$ is a solution of $x^2 - dy^2 = 1$. For each positive integer n, let (x_n, y_n) be determined by $x_n + y_n\sqrt{d} = (u + v\sqrt{d})^n$. The numbers x_n and y_n can be determined recursively by

$$x_{n+1} = ux_n + dvy_n,$$

$$y_{n+1} = vx_n + uy_n.$$

However, it is possible to derive expressions for x_n and y_n as polynomials in u and v.

Exercise 3.1. Using the fact that $dv^2 = u^2 - 1$, verify that

$$\begin{aligned}(x_1, y_1) &= (u, v),\\(x_2, y_2) &= (2u^2 - 1, 2uv),\\(x_3, y_3) &= (4u^3 - 3u, (4u^2 - 1)v),\\(x_4, y_4) &= (8u^4 - 8u^2 + 1, (8u^3 - 4u)v).\end{aligned}$$

Exercise 3.2. Prove that (x_n, y_n) can be written in the form

$$(T_n(u), U_n(u)v),$$

where $T_n(u)$ and $U_n(u)$ are both polynomials in u of respective degrees n and $n - 1$. Determine a recursion for the sequence of pairs $(T_n(u), U_n(u))$.

Exercise 3.3. Make a list of the functions $T_n(u)$ and $U_n(u)$. Each of the sequences individually satisfies a recursion that does not involve entries in the other sequence. Make a conjecture and prove it.

Exercise 3.4. Express $\cos 2\theta$, $\cos 3\theta$, and $\cos 4\theta$ as polynomials in $\cos\theta$.

Exercise 3.5. Show that for each positive integer n, $\sin n\theta$ can be written as the product of $\sin\theta$ and a polynomial in $\cos\theta$. Determine what this polynomial is when $n = 2, 3, 4$.

Exercise 3.6. Compare the polynomials obtained in Exercises 3.4 and 3.5 with the polynomials $T_n(u)$ and $U_n(u)$.

Exercise 3.7. As in Exercise 1.2.3, it is posible to use matrix theory to derive recursions for the sequences $\{x_n\}$ and $\{y_n\}$. Observe that

$$\begin{pmatrix} x_{n+1} \\ y_{n+1} \end{pmatrix} = \begin{pmatrix} u & dv \\ v & u \end{pmatrix} \begin{pmatrix} x_n \\ y_n \end{pmatrix}.$$

Verify that

$$\begin{pmatrix} u & dv \\ v & u \end{pmatrix}^2 = 2u \begin{pmatrix} u & dv \\ v & u \end{pmatrix} - \begin{pmatrix} 1 & 0 \\ 0 & 1 \end{pmatrix}$$

and deduce that

$$x_{n+1} = 2ux_n - x_{n-1},$$
$$y_{n+1} = 2uy_n - y_{n-1},$$

for $n \geq 1$, where $(x_0, y_0) = (1, 0)$ and $(x_1, y_1) = (u, v)$.

Exploration 3.3. Suppose $u^2 - dv^2 = -1$. Then if $x_n + y_n\sqrt{d} = (u + v\sqrt{d})^n$, then $x_n^2 - dy_n^2 = (-1)^n$. Investigate expressions for x_n and y_n in terms of u and v. What recursions are satisfied by the sequences $\{x_n\}$ and $\{y_n\}$.

3.4 Chebyshev Polynomials

The polynomials of Section 3.3 are well known. The T_n are called *Chebyshev polynomials* (or *Chebyshev polynomials of the first kind*), and the U_n are called *Chebyshev polynomials of the second kind.* They turn up in a variety of mathematical contexts and have a number of remarkable properties. We look at a few of them here. In this section we will define them anew and show that our definition is consistent with that of Section 3.3.

For $-1 < t < 1$, determine θ such that $0 < \theta < \pi$ and $t = \cos\theta$ (i.e., $\theta = \arccos t$). Define

$$T_n(t) = \cos n\theta = \cos(n \arccos t)$$

and

$$U_n(t) = \frac{\sin n\theta}{\sin\theta} = (1 - t^2)^{-\frac{1}{2}} \sin(n \arccos t).$$

While these functions are initially defined on a restricted domain, they turn out to be polynomials in t and so have meaning for all real values of t.

Exercise 4.1. Consider the equation $x^2 - (t^2 - 1)y^2 = 1$ where t is a parameter. An obvious solution is $(x, y) = (t, 1)$. Other solutions can be obtained from

$$x_n + \sqrt{t^2 - 1}y_n = \left(t + \sqrt{t^2 - 1}\right)^n = \left(t + i\sqrt{1 - t^2}\right)^n.$$

Writing $t = \cos\theta$ and using de Moivre's theorem, verify that

$$(x_n, y_n) = (T_n(t), U_n(t)).$$

Exercise 4.2.

(a) Verify that $T_0(t) = 1$, $T_1(t) = t$ and establish the recursion, for $n \geq 2$,

$$T_n(t) = 2tT_{n-1}(t) - T_{n-2}(t).$$

(b) Use (a) to determine $T_2(t)$, $T_3(t)$, and $T_4(t)$.

(c) Through the substitution $t = \cos\theta$, check your answers for (b) against your answers for Exercise 3.4.

(d) Prove that $T_n(t)$ is a polynomial of degree n with integer coefficients, leading coefficient 2^n, and n distinct real roots lying in the closed interval $[-1, 1]$.

(e) Let k be a nonnegative integer. Prove that $T_{2k+1}(0) = 0$ and $T_{2k}(0) = (-1)^k$.

Exercise 4.3.

(a) Verify that $U_0(t) = 0$, $U_1(t) = 1$ and establish the recursion, for $n \geq 2$,

$$U_n(t) = 2tU_{n-1}(t) - U_{n-2}(t).$$

(b) Use (a) to determine $U_2(t)$, $U_3(t)$, and $U_4(t)$.

(c) Compare your answers for (b) with your answers to Exercise 3.1.

(d) Prove that $U_n(1) = n$ for each positive integer n.

(e) What is the degree of $U_n(t)$? Its leading coefficient? Discuss the roots of $U_n(t)$.

Exercise 4.4.

(a) From the expansions of $\cos(n + 1)\theta$ and $\sin(n + 1)\theta$, obtain the equations

$$T_{n+1}(t) = tT_n(t) + (t^2 - 1)U_n(t),$$
$$U_{n+1}(t) = T_n(t) + tU_n(t),$$

for $n \geq 0$.

Exercise 4.5. Prove that $\frac{d}{dt}T_n(t) = nU_n(t)$ for $n \geq 1$.

Exercise 4.6.

(a) Prove that $T_{2n}(t) = 1 + 2(t^2 - 1)U_n^2(t)$ for $n \geq 0$.

(b) Prove that $U_{2n}(t) = 2T_n(t)U_n(t)$ for $n \geq 0$.

Exercise 4.7. Prove that $T_{2n+1}(t) = 1 + (t - 1)(U_{n+1}(t) + U_n(t))^2$ for $n \geq 0$.

Exercise 4.8. For positive integers m, n, prove that:
(a) $T_{mn}(t) = T_m(T_n(t))$.

Exercise 4.9. Suppose that w and z are indeterminates related by the equation

$$w = (z^2 - 1)^{\frac{1}{2}}.$$

(a) Expand each of the functions $(z + w)^2$, $(z + w)^2$, $(z + w)^3$, and $(z + w)^4$ in the form $p(z) + q(z)w$, where $p(z)$ and $q(z)$ are polynomials.
(b) Make a conjecture, and prove it, about a similar expansion for $(z + w)^n$.

Exercise 4.10. Using de Moivre's theorem and the fact that for $t = \cos\theta$,

$$T_n(t) + i(1 - t^2)^{\frac{1}{2}} U_n(t) = \cos n\theta + i \sin n\theta,$$

provide a determination of $T_n(T)$ and $U_n(t)$ as polynomials in t that will lead to an alternative proof of the result of Exercise 4.9.

Exercise 4.11. Prove that for d a positive nonsquare integer and n a positive integer,

$$\left(u + v\sqrt{d}\right)^n = T_n(u) + vU_n(t)\sqrt{d}$$

whenever $u^2 - dv^2 = 1$. Compare the results of Section 3.♠

In the next two exercises we require some terminology that will be reviewed in detail in Section 4.1. We say that $a \equiv b$ modulo m if $a - b$ is a multiple of m, for integers a, b, m. Each number is congruent, modulo m, to its remainder upon division by m, that is, to one of the numbers $0, 1, 2, \ldots, m - 1$.

Exercise 4.12. Let $\{a_n\}$ be a sequence of integers defined for $n = 0, \pm 1, \pm 2, \ldots$ that satisfies a recursion of the type

$$a_{n+1} = 2ca_n - a_{n-1}$$

where c is a constant integer. Observe that the recursion is symmetric in that given two consecutive terms of the sequence, we can "go backward" by the same rule:

$$a_{n-1} = 2ca_n - a_{n+1}.$$

Let m be a positive integer exceeding 1, so that modulo m, each term is congruent to one of $0, 1, 2, \ldots, m - 1$.
(a) Argue that there are only finitely many incongruent pairs (a_n, a_{n+1}) of consecutive terms, modulo m.
(b) Prove that there are integers r and s for which $r < s$ and $a_r \equiv a_s$, $a_{r+1} \equiv a_{s+1}$.
(c) Deduce that $a_0 = a_{s-r}$ and $a_1 = a_{s-r+1}$, so that $\{a_n\}$ is periodic, modulo m.

Exercise 4.13. Suppose that u and m are positive integers. Prove that there is a positive integer n such that $T_n(u) \equiv 1$ and $U_n(u) \equiv 0$, modulo m.

Exercise 4.14. Let m be an integer exceeding 1, and let S_m be the set of all natural numbers n for which both $(m^2 - 1)n + 1$ and $m^2 n + 1$ are perfect squares. For example, S_2 contains the number 56. Prove that S_m is a nonempty set and determine the greatest common divisor of the numbers in S_m. (Problem 10879 in the *American Mathematical Monthly* 108 (2001), 565.)

3.5 Related Pell's Equations

While getting a starting solution will be dealt with in Chapter 4, the exercises in this section will indicate how we can find solutions when the parameter is simply related to other parameters for which a solution is known.

Exercise 5.1. Let d be a nonsquare integer whose largest square divisor is m, so that $d = m^2 e$ for some square-free integer e.

(a) Prove that a Pell's equation $x^2 - dy^2 = k$ is solvable if and only if $x^2 - ey^2 = k$ is solvable.

(b) Given a solution of $x^2 - ey^2 = 1$, explain how to find a solution of $x^2 - dy^2 = 1$.

Exercise 5.2. From the smallest positive solution $(x, y) = (3, 2)$ of $x^2 - 2y^2 = 1$, determine the smallest positive solution of $x^2 - dy^2 = 1$ for $d = 8$, 18, 32, 50, 72, 98, 128, 162, 200.

Exercise 5.3. From the solution of $(x, y) = (2, 1)$ of $x^2 - 3y^2 = 1$, determine the solutions of $x^2 - dy^2 = 1$ for $d = 12$, 27, 48, 75, 108.

3.6 Notes

1.9. The two equations in the system imply that

$$(x^2 - 1)(y^2 - 1) = (u^2 - 1)(v^2 - 1).$$

Note that every solution of this equation is a solution of the original system. However, this equation, but not the system, is satisfied by

$$(x, y, u, v) = (F_{2n-1}, F_{2n-3}, F_{2n}, F_{2n-2}),$$

where $\{F_n\}$ is the Fibonacci sequence ($F_0 = F_1 = 0$ and $F_{n+1} = F_n + F_{n-1}$ for each n). This is a consquence of the general equations $F_{n+2} = 3F_n - F_{n-2}$, $F_{2n}^2 - 1 = F_{2n-2}F_{2n+2}$ and $F_{2n-1}^2 - 1 = F_{2n+2}F_{2n}$.

3.7 Hints

1.7. Note that $(x, y) = (3, 2)$ satisfies $x^2 - 2y^2 = 1$ and make repeated use of Exercise 1.4.

1.9(a). Multiply the first equation by $v + 1$ and the second by $v - 1$ and subtract.

1.10(a). Look at $m = 1, 2, 3, 4$. It is not hard to find suitable values of r (not exceeding 100); now guess what a general formula for r in terms of m might be.

2.1. Note that $N\left(c + \sqrt{d}\right) = k$. For (b), divide the equation $x^2 - dy^2 = r^2$ by r^2.

2.2(a). To narrow the search, observe that $13y^2 = (x + 3)(x - 3)$, so that 13 must divide either $x - 3$ or $x + 3$. Use this factorization to find a solution without recourse to a calculator.

2.2(d). Multiply numerator and denominator by $11 + 3\sqrt{13}$ (why?).

4

The Fundamental Solution

4.1 Existence of a Solution in Integers

In Section 3.2, we saw how, using various tricks, solutions in rational x and y of $x^2 - dy^2 = 1$ could be obtained from two solutions of an equation $x^2 - dy^2 = k$. Sometimes, the rational numbers turned out to be integers. The chances of this happening would apparently improve with the number of solutions of $x^2 - dy^2 = k$ for a particular k. This suggests that it might be useful to look for values of k for which there are a lot of solutions. This strategy succeeds spectacularly; there are infinitely many solutions for suitable k.

Before turning to the exercises, there is a little background to be reviewed. Some of the exercises depend on the application of the *pigeonhole principle*, whose first explicit application was made in 1842 in a number theory context by Gustav Peter Lejeune Dirichlet (1805–1859). (The reader may be interested in learning that, in 1831, he married Rebecca Mendelssohn, a sister of the famous composer Felix Mendelssohn.) This innocuous-sounding but highly useful principle states that if we have to sort n objects (pigeons) among m categories (the pigeonholes) and n exceeds m, then there must be two objects that fall into the same category.

In some of the exercises it is helpful to have access to the notion of modular arithmetic. Let a, b, and m be integers, with m nonzero. We say that $a \equiv b \pmod{m}$ (read *a is congruent to b modulo m*) if $a - b$ is a multiple of m. Another way of putting it is that m is a divisor of $a - b$; if we divide m into each of a and b, obtaining nonnegative remainders less than m, then these remainders must be equal.

In what follows, d is a nonsquare positive integer and k is an integer.

Exercise 1.1. Verify the following:
(a) $17 \equiv 35 \pmod{6}$.
(b) $-13 \equiv 33 \pmod{23}$.
(c) Every even number is congruent to 0 modulo 2.

Exercise 1.2. Suppose that $a \equiv b \pmod{m}$ and $c \equiv d \pmod{m}$. Show that

$$a + c \equiv b + d \pmod{m},$$
$$ac \equiv bd \pmod{m}.$$

Exercise 1.3. Suppose that (x_1, y_1) and (x_2, y_2) are solutions in integers of the equation $x^2 - dy^2 = k$ for which

$$x_1 \equiv x_2 \pmod{k} \quad \text{and} \quad y_1 \equiv y_2 \pmod{k}.$$

Verify that:

(a) $x_1x_2 - dy_1y_2 \equiv x_1^2 - dy_1^2 \equiv 0 \pmod{k}$.
(b) $x_1y_2 - x_2y_1 \equiv 0 \pmod{k}$.
(c) $(x, y) = (x_1x_2 - dy_1y_2, x_1y_2 - x_2y_1)$ is a solution of $x^2 - dy^2 = k^2$.
(d) $x^2 - dy^2 = 1$ has an integer solution.

Exercise 1.4. Suppose that $m \geq k^2 + 1$ and that $(x, y) = (x_1, y_1), (x_2, y_2), \dots, (x_m, y_m)$ are solutions in integers of the equation $x^2 - dy^2 = k$. Prove that there exist solutions (x_i, y_i) and (x_j, y_j) for which $1 \leq i, j \leq m$, $x_i \equiv x_j$, and $y_i \equiv y_j \pmod{k}$. ♠

For any real number α, denote by $\lfloor\alpha\rfloor$ the largest integer that does not exceed α and by $\langle\alpha\rangle$ the *fractional part* $\alpha - \lfloor\alpha\rfloor$ of α. Observe that $0 \leq \langle\alpha\rangle < 1$.

Exercise 1.5. Consider the numbers $\langle\sqrt{3}\rangle, \langle 2\sqrt{3}\rangle, \dots, \langle 10\sqrt{3}\rangle$, and $\langle 11\sqrt{3}\rangle$.

(a) Why would you expect to find that two of these eleven numbers have the same first digit after the decimal point?
(b) Indeed, verify that $\langle 3\sqrt{3}\rangle = 3\sqrt{3} - 5 = 0.196152\dots$ and $\langle 7\sqrt{3}\rangle = 7\sqrt{3} - 12 = 0.124355\dots$.
(c) From the two equations in (b), deduce that $|7 - 4\sqrt{3}| = 0.071\dots < \frac{1}{10} < \frac{1}{4}$.

Exercise 1.6. Let N be a given integer. We will show that there are positive integers u and v for which

$$|u - v\sqrt{d}| < \frac{1}{N} \leq \frac{1}{v}.$$

(a) Consider the $N+1$ numbers $\langle\sqrt{d}\rangle, \langle 2\sqrt{d}\rangle, \dots, \langle N\sqrt{d}\rangle, \langle(N+1)\sqrt{d}\rangle$. Show that for some value of i with $0 \leq i \leq N - 1$, two of these numbers, say $\langle q\sqrt{d}\rangle$ and $\langle s\sqrt{d}\rangle$, must fall in the interval $\{t : i/N < t < (i + 1)/N\}$. We may suppose that $q > s$.
(b) Deduce from (a) that there are positive integers p, q, r, s for which

$$\frac{i}{N} < q\sqrt{d} - p < \frac{i+1}{N} \quad \text{and} \quad \frac{i}{N} < s\sqrt{d} - r < \frac{i+1}{N}.$$

Verify that $p = \lfloor q\sqrt{d}\rfloor$, $r = \lfloor s\sqrt{d}\rfloor$, and that $p > r$.

(c) Show that $|q - s| \leq N$ and $|(p - r) - (q - s)\sqrt{d}| < 1/N$, and deduce the existence of numbers u and v as specified.

(d) By noting that $u + v\sqrt{d} = (u - v\sqrt{d}) + 2v\sqrt{d}$, prove that

$$|u^2 - dv^2| \leq 2\sqrt{d} + 1.$$

Exercise 1.7. Show that there are infinitely many pairs (u, v) of integers for which $|u - v\sqrt{d}| < 1/v$.

Exercise 1.8. Use the results of Exercises 1.6 and 1.7 to conclude that there is an integer k not exceeding $2\sqrt{d} + 1$ in absolute value for which $x^2 - dy^2 = k$ has infinitely many solutions.

Exercise 1.9. Use the results of Exercises 1.8, 1.4, and 1.3 to conclude that $x^2 - dy^2 = 1$ has a solutions in positive integers x and y. ♠

The following argument for deducing the existence of an integer solution to $x^2 - dy^2 = 1$ in the event that $x^2 - dy^2 = k$, for k prime, has two solutions, is due to J.L. Lagrange. The ideas behind this argument were extended by him to the case that k is a product of distinct primes, then of prime powers.

Exercise 1.10. Suppose that $u^2 - dv^2 = w^2 - dz^2 = k$, where k is prime.

(a) Verify that

$$(uw \pm dvz)^2 - d(uz \pm vw)^2 = k^2$$

and that

$$k(z^2 - v^2) = (uz + vw)(uz - vw).$$

(b) Deduce from (a) that k must divide $uw + dvz$ or $uw - dvz$ and thence obtain a solution in integers to $x^2 - dy^2 = 1$.

(c) Adapt the foregoing argument to the case that k is a product of two or more distinct primes.

(d) Adapt the foregoing argument to the case that k is a power of a prime.

4.2 The Fundamental Solution

When d is a nonsquare positive integer, the equation $x^2 - dy^2 = 1$ is solvable in integers. More can be said: There is a fundamental solution from which every other solution can be obtained. Specifically, there are positive integers x_1 and y_1 such that

$$\{(\pm x_n, y_n) : x_n \text{ and } y_n \text{ are integers};$$
$$x_n + y_n\sqrt{d} = (x_1 + y_1\sqrt{d})^n;\ n = 0, \pm 1, \pm 2, \ldots\}$$

yields a complete set of solutions.

Exercise 2.1. Let $x_1 > 1$, $y_1 \geq 1$, and $(x_n + y_n\sqrt{d}) = (x_1 + y_1\sqrt{d})^n$. Show that when n is a positive integer, then $x_{n+1} > x_n$ and $y_{n+1} > y_n$. ♠

Exercise 2.1 suggests that if (x_1, y_1) is to do the job required of it, then these numbers should be as small as possible. We select x_1 and y_1 so that both are positive and x_1 is the smallest positive integer satistying the equation $x_1^2 - dy_1^2 = 1$.

Exercise 2.2. Show that if p, q, r, s are positive integers for which $p > r$ and $p^2 - dq^2 = r^2 - ds^2 = 1$, then $q > s$ and $p + q\sqrt{d} > r + s\sqrt{d}$.

Exercise 2.3. Suppose that $u > 0$, $v > 0$, $u^2 - dv^2 = 1$.

(a) Argue from the definition of (x_1, y_1) that

$$u + v\sqrt{d} \geq (x_1 + y_1\sqrt{d}).$$

(b) Define the positive integer m by the condition

$$(x_1 + y_1\sqrt{d})^m \leq (u + v\sqrt{d}) < (x_1 + y_1\sqrt{d})^{m+1}.$$

Deduce that

$$1 \leq (u+v\sqrt{d})(x_1+y_1\sqrt{d})^{-m} = (u+v\sqrt{d})(x_1-y_1\sqrt{d})^{+m} < (x_1+y_1\sqrt{d}).$$

(c) Suppose the integers a and b are determined by

$$a + b\sqrt{d} = (u + v\sqrt{d})(x_1 - y_1\sqrt{d})^m.$$

Observe that

$$a^2 - db^2 = 1,$$
$$a - b\sqrt{d} = (a + b\sqrt{d})^{-1} > 0,$$

and

$$a - b\sqrt{d} \leq 1 \leq a + b\sqrt{d}.$$

(d) Deduce that $a > 0$.

(e) From the minimality of (x_1, y_1) and Exercise 2.2, conclude that $b = 0$ and $a = 1$, so that $u + v\sqrt{d} = (x_1 + y_1\sqrt{d})^m$.

Exercise 2.4. Suppose that the integer pair (u, v) is an arbitrary solution of $x^2 - dy^2 = 1$. Deduce from Exercise 2.3 that:

(a) $|u| + |v|\sqrt{d} = (x_1 + y_1\sqrt{d})^m$ for some nonnegative integer m.
(b) $|u| - |v|\sqrt{d} = (x_1 + y_1\sqrt{d})^m$ for some negative integer m.
(c) Either (u, v) or $(-u, v)$ is of the form (x_m, y_m) for some integer m.

Exercise 2.5. Suppose that $x^2 - dy^2 = -1$ is solvable and that the solution with smallest positive integers is $(x, y) = (r_1, s_1)$. With (x_1, y_1) as the fundamental solution of $x^2 - dy^2 = 1$, prove that:

(a) $x_1 + y_1\sqrt{d} = (r_1 + s_1\sqrt{d})^2$.

(b) Every solution (x, y) of $x^2 - dy^2 = -1$ in positive integers has the form (r_n, s_n), where n is odd and $r_n + s_n\sqrt{d} = (r_1 + s_1\sqrt{d})^n$.

Exercise 2.6. Newton's algorithm for approximating the square root of a positive integer d can be formulated as follows. Suppose you begin with the positive approximation r. Then for the next approximation, take the average of r and d/r, namely $\frac{1}{2}(r + d/r)$. Observe that the average is taken of one number less than and one number greater than $\sqrt{d}$. Repeat this process to get a sequence of approximations.

(a) Prove that $\sqrt{d}$ lies between r and d/r, and that the average $\frac{1}{2}(r + d/r)$ is greater than $\sqrt{d}$ and is closer to $\sqrt{d}$ than one of r and d/r.

(b) Suppose that the first guess is the rational number x_1/y_1, where (x_1, y_1) is the fundamental solution of $x^2 - dy^2 = 1$. Describe the sequence of approximants obtained by iterating Newton's method.

Exercise 2.7. Let $\{a_n\}$ be a sequence of integers defined for $n = 0, \pm 1, \pm 2, \ldots$ that satisfies a recursion of the type

$$a_{n+1} = 2ca_n - a_{n-1},$$

where c is a constant integer. Observe that the recursion is symmetric in that given two consecutive terms of the sequence, we can "go backward" by the same rule:

$$a_{n-1} = 2ca_n - a_n.$$

Let m be a positive integer exceeding 1, so that, modulo m, each term is congruent to one of $0, 1, 2, \ldots, m-1$.

(a) Argue that there are only finitely many incongruent pairs (a_n, a_{n+1}) of consecutive terms, modulo m.

(b) Prove that there are integers r and s for which $r < s$ and $a_r \equiv a_s, a_{r+1} \equiv a_{s+1}$ (mod m).

(c) Deduce that $a_0 \equiv a_{s-r}$ and $a_1 \equiv a_{s-r+1}$, so that $\{a_n\}$ is periodic, modulo m.

♠

3.3 Algebraic Integers

There is a more natural setting in which to consider solutions of Pell's equation. The set $\mathbf{Q}(\sqrt{d})$ of numbers of the form $r + s\sqrt{d}$, with r and s rational, is closed under the arithmetic operations of addition, subtraction, multiplication, and division by nonzero numbers. The integers consitute a special subset of the ordinary rationals, and we seek a subset of this larger family $\mathbf{Q}(\sqrt{d})$ of numbers that generalizes the integers.

We will suppose that d is *square-free*. This means that d is a product of distinct primes, so that it is not evenly divisible by any square number other than 1.

Exercise 3.1. How much of a loss of generality is imposed by the condition that d be square-free?

Exercise 3.2.

(a) Consider the quadratic equation $t^2 + bt + c = 0$, where b and c are integers. Suppose that it has a rational root r; prove that this root must in fact be an integer.

(b) Show that all of the *rational* roots of a polynomial equation $t^n + a_{n-1}t^{n-1} + a_{n-2}t^{n-2} + \cdots + a_1 t + a_0 = 0$ with integer coefficients are in fact integers.

♠

Taking our cue from Exercise 3.1, we say that a number θ is a *quadratic integer* if it is a root of a monic quadratic equation $t^2 + bt + c = 0$ with integer coefficients. More generally, θ is an *algebraic integer* if it is a root of a monic polynomial equation $t^n + a_{n-1}t^{n-1} + a_{n-2}t^{n-2} + \cdots + a_1 t + a_0 = 0$ with integer coefficients. A *unit* is an algebraic integer whose reciprocal is also an algebraic integer.

Exercise 3.3. By finding a monic quadratic equation with integer coefficients of which it is a root, verify that $u + v\sqrt{d}$ is an algebraic integer whenever u and v are integers.

Exercise 3.4. Let b and c be integers and u and v be rationals with $v \neq 0$. Suppose that $u + v\sqrt{d}$ satisfies the equation $t^2 + bt + c = 0$.

(a) Prove that $2u$ must be an integer and that $4dv^2 = b^2 - 4c$.

(b) Prove that $4v^2$, and hence $2v$, is an integer.

(c) If b is odd, prove that $d \equiv 1 \pmod 4$ and that $2v$ is an odd integer.

Exercise 3.5. Using Exercise 3.4, deduce the following result:

If d is a squarefree integer, then $u + v\sqrt{d}$ is a quadratic integer if and only if either

(a) $d \not\equiv 1 \pmod 4$ and u and v are integers or

(b) $d \equiv 1 \pmod 4$ and $u = p/2$ and $v = q/2$, where p and q are integers with the same parity.

Exercise 3.6. Show that a quadratic integer $r + s\sqrt{d}$ is a unit if and only if its norm $r^2 - ds^2$ is equal to ± 1.

Exercise 3.7. Let $d \equiv 1 \pmod 4$. Show that the solutions of $x^2 - dy^2 = 1$, where $x + y\sqrt{d}$ is a quadratic integer, can be obtained from the solutions of $x^2 - dy^2 = 4$, for which x and y have the same parity.

Exercise 3.8. Show that if $x^2 - dy^2 = 4$ can be solved for *odd* integers x and y, then $d \equiv 5 \pmod 8$.

Exercise 3.9. Consider values of d congruent to 5 modulo 8, i.e., of the form $8k + 5$. For each value $d = 5, 13, 21, 29, 37, 45, 53, \ldots$, try to determine the smallest positive solution of $x^2 - dy^2 = 4$ in odd integers x and y. By considering the positive powers of $x + y\sqrt{d}$, determine other solutions of the same equation. Can you derive solutions of the equation $x^2 - dy^2 = 1$ from any of these solutions?

Exercise 3.10. For $d \equiv 5 \pmod 8$, modify the proof in Section 4.2 to obtain the existence of an algebraic integer $u + v\sqrt{d}$ with $u \geq 0$ such that its integer powers constitute exactly the set of algebraic integers $x + y\sqrt{d}$ with norm 1 and $x > 0$.

Exercise 3.11. Suppose that m is an odd positive integer and that $d = m^2 - 4$.
(a) Verify that d is congruent to 5 modulo 8, and determine two solutions for $x^2 - dy^2 = 4$ for which x and y are positive odd integers.
(b) Prove that d must be divisible by a prime congruent to 3 modulo 4, and deduce that $x^2 - dy^2 = -4$ has no solutions in integers x and y.

Exercise 3.12. Suppose that m is an odd positive integer and that $d = m^2 + 4$.

Verify that d is congruent to 5 modulo 8, and determine solutions in positive odd integers to each of the equations

$$x^2 - dy^2 = -4,$$
$$x^2 - dy^2 = 4.$$

Exercise 3.13. Suppose that $d \equiv 5 \pmod 8$, and let p and q be odd integers for which $p^2 - dq^2 = 4$. This will yield a rational solution $(x, y) = \left(\frac{p}{2}, \frac{q}{2}\right)$ for $x^2 - dy^2 = 1$. Prove that if $u + v\sqrt{d} = \left(\frac{p}{2} + \frac{q}{2}\sqrt{d}\right)^3$, then $(x, y) = (u, v)$ is a solution of $x^2 - dy^2 = 1$ in positive integers.

4.4 A Bilateral Sequence

In this section we examine solutions of $x^2 - dy^2 = k$. In particular, we find that if there is a solution to this equation, then we can find one for which x and y do not exceed certain computable bounds that depend on k and the fundamental solution of $x^2 - ky^2 = 1$.

Suppose that $(x, y) = (u, v)$ is the fundamental solution of $x^2 - dy^2 = 1$. Let $(x, y) = (r, s)$ be a solution of $x^2 - dy^2 = k$; then $(x, y) = (ru + dsv, rv + su)$ is another solution. Suppose that $z = rv + su$. We form a symmetric quadratic equation involving s and z that is the tool for defining a bilateral sequence.

Exercise 4.1.
(a) By setting $r = (z - su)/v$, obtain

$$(z - su)^2 - dv^2s^2 = v^2k$$

and deduce that

$$z^2 - 2usz + s^2 - v^2k = 0. ♠$$

Let $p(y, z) = y^2 - 2uyz + z^2 - v^2k$. Observe that $p(y, z)$ is symmetric in y and z. We define a sequence $\{s_n\}$ as follows. Let $s_0 = s$. Let s_{-1} and s_1 be the two solutions of $p(s_0, z) = 0$ with $s_{-1} \leq s_1$. Suppose that s_i has been defined for $|i| \leq n$, so that $p(s_i, s_{i+1}) = 0$ for $-n \leq i \leq n-1$. Verify that s_{n-1} is a root of the equation $p(s_n, z) = 0$. Define s_{n+1} to be the second root. Verify that s_{-n+1} is a root of the equation $p(s_{-n}, z) = 0$. Define s_{-n-1} to be the second root of the equation.

Exercise 4.2.

(a) Verify that $\{s_n\}$ is a sequence of integers for which $ds_n^2 + k$ is always an integer square.

(b) Prove that $\{s_n\}$ satisfies the recursions

$$\begin{aligned} s_{n+1} + s_{n-1} &= 2us_n, \\ s_{n+1}s_{n-1} &= s_n^2 - v^2k. \end{aligned}$$

Exercise 4.3. Let $k < 0$ and suppose a sequence $\{s_n\}$ has been defined as in Exercise 5.1.

(a) Suppose that $\{s_n\}$ contains a positive integer. Prove that $\{s_n\}$ consists entirely of positive integers. We will suppose that s_0 is the smallest term in the sequence, and that $s_{-1} \leq s_1$.

(b) Prove that

$$\frac{s_{n+1}}{s_n} > \frac{s_n}{s_{n-1}}$$

and deduce that $s_{n+1} > s_n$ when $n \geq 0$ and $s_{n+1} \leq s_n$ when $n \leq -1$.

(c) Prove that when $n \geq 0$,

$$\begin{aligned} s_{n+1} &= us_n + v\sqrt{ds_n^2 + k}, \\ s_{n-1} &= us_n - v\sqrt{ds_n^2 + k}, \end{aligned}$$

while if $n < 0$,

$$\begin{aligned} s_{n+1} &= us_n - v\sqrt{ds_n^2 + k}, \\ s_{n-1} &= us_n + v\sqrt{ds_n^2 + k}. \end{aligned}$$

Exercise 4.4. Let $k < 0$ and $\{s_n\}$ be as above.

(a) Prove that

$$\left| v\sqrt{ds_0^2 + k} \right| \leq (u-1)s_0.$$

(b) Obtain the inequality

$$s_0 \leq v\sqrt{\frac{|k|}{2(u-1)}}.$$

Exercise 4.5.

(a) Prove that $p(s_n, z) = 0$ can have a double root if and only if $k = -dm^2$ for some m.

(b) If $k = -dm^2$ and $s_0 = m$, prove that $s_{-n} = s_n$ for each integer n.

(c) Prove that $x^2 - dy^2 = -dm^2$ if $x = dz$ for an integer satisfying $y^2 - dz^2 = m^2$.

Exercise 4.6. Suppose that $k > 0$ and that a sequence $\{s_n\}$ has been defined as in Exercise 5.1 and contains at least one nonnegative term. Suppose that the indexing of the sequence is such that s_0 is the smallest nonnegative term.

(a) Prove that for any real y, $p(y, y) < 0$, and deduce that y must lie strictly between the two solutions z of the equation $p(y, z) = 0$.

(b) If $s_{-1} < s_0 < s_1$, prove that $\{s_n\}$ is an increasing sequence and that it contains both positive and negative terms.

(c) Prove that $s_0^2 < v^2 k$, and so deduce that if $x^2 - dy^2 = k$ has a solution, then there is at least one solution for which $0 \le y < v\sqrt{k}$, $0 \le |x| < u\sqrt{k}$.

(d) Prove that when $n \ne 0, -1$, then $s_n^2 \ge v^2 k$.

Exercise 4.7. We can sharpen the bounds on the smallest solutions of $x^2 - dy^2 = k$ when $k > 0$. Consider the positive solution $(x, y) = (r, s)$ with r assuming the smallest positive value of x. (This can occur when $s = s_0$ or $s = s_{-1}$ in the notation of Exercise 4.6.)

Recall that $(r, s) * (u, -v) = (ur - dvs, us - vr)$.

(a) Prove that $d^2v^2s^2 < u^2r^2$, so that $ur > dvs$.

(b) Deduce that $ur - dvs \ge r$, whence

$$r^2(u - 1) \ge (u + 1)(r^2 - k).$$

(c) Conclude that $2r^2 \le k(u + 1)$.

(d) Prove that

$$s^2 \le \frac{kv^2}{2(u + 1)}.$$

(e) Deduce that either s_0 or $|s_{-1}|$ does not exceed $v\sqrt{k/(2(u + 1))}$, so that $x^2 - dy^2 = k$ has a solution for which

$$|x| \le \sqrt{\frac{k(u + 1)}{2}} \quad \text{and} \quad |y| \le v\sqrt{k/(2(u + 1))}.$$

Exercise 4.8. Consider the equation $x^2 - 2y^2 = 7$. This has solution $(x, y) = (3, 1)$. Follow the process of Exercise 4.1 with $s_0 = 1$ to obtain the sequence

$$\{\ldots, -3771, -647, -111, -19, -3, 1, 9, 53, 309, 1801, 10497, \ldots\}.$$

Check that it satisfies the recursions of Exercise 4.2(b) and yields solutions to $x^2 - 2y^2 = 7$.

Exercise 4.9. Consider the equation $x^2 - 2y^2 = -7$. This has solution $(x, y) = (1, 2)$. Follow the process of Exercise 5.1 with $s_0 = 2$ to obtain the sequence

$$\{\ldots, 4348, 746, 128, 22, 4, 2, 8, 46, 268, 1562, \ldots\}.$$

Exercise 4.10. Study the solutions of the following equations: $x^2 - 2y^2 = k$ with $k = -2, -8, 4, 1022$.

4.5 Explorations

Exploration 4.1. *The Legendre symbol.* If the equation $x^2 - dy^2 = k$ is solvable in integers, then in particular $x^2 \equiv k \pmod{d}$, so that k must be a square modulo d, and a fortiori a square modulo p for any prime divisor p of d. We define the *Legendre symbol* $\left(\frac{k}{p}\right)$ by

$$\left(\frac{k}{p}\right) = \begin{cases} 1 & \text{if } k \text{ is not a multiple of } p \text{ and } k \equiv m^2 \pmod{p} \\ & \quad \text{for some integer } m; \\ -1 & \text{if } k \not\equiv m^2 \pmod{p} \text{ for each integer } m; \\ 0 & \text{if } k \text{ is a multiple of } p. \end{cases}$$

For example, regardless of the prime p, $\left(\frac{1}{p}\right)$ and $\left(\frac{4}{p}\right)$ are both equal to 1. What can be said about $\left(\frac{-1}{p}\right)$ and $\left(\frac{2}{p}\right)$? Is $\left(\frac{uv}{p}\right) = \left(\frac{u}{p}\right)\left(\frac{v}{p}\right)$?

Exploration 4.2. Examine the sequence of solutions of $x^2 - 5y^2 = \pm 4$. What is interesting about the values of x and y that appear?

Exploration 4.3. In Exercise 3.8 it was found that a necessary condition on d for $x^2 - dy^2 = 4$ to have a solution (x, y) in odd integers was that $d \equiv 5 \pmod{8}$. Is this condition also sufficient? Can you determine other conditions on d that will provide for such a solution?

Exploration 4.4. For each prime number p, examine the structure of the set N_p of numbers of the form $x^2 - py^2$, where x and y range over the integers. This set is closed under multiplication; is there a set of "prime" elements such that every number in N_p can be uniquely representable as a product of these "primes"? How about the ordinary primes?

Exploration 4.5. *Quadratic forms.* The function $x^2 - dy^2$ is a homogeneous polynomial of degree 2, and as such is a quadratic form in two variables. The general quadratic form is given by $ax^2 + hxy + by^2$, where a, b, and h are integers. Other examples are $x^2 - y^2$, $x^2 + y^2$, $x^2 + xy + y^2$, and $x^2 - xy + y^2$. Given a quadratic form, what can be said about the set of values assumed by the form when x and y are integers?

It is particularly easy to characterize the numbers that can be written in the form $x^2 - y^2 = (x+y)(x-y)$. Any such number must be expressible as the product of two integers of the same parity. Is the converse true? Is the set of numbers of the form $x^2 - y^2$ closed under multiplication?

The numbers of the form x^2+y^2 present more of a challenge. If we let $z = x+yi$ and $w = u+vi$, then using the fact that $|zw| = |z||w|$, one can express the product $(x^2+y^2)(u^2+v^2)$ as a sum of two squares. Thus, the collection of numbers so representable is closed under multiplication. What numbers are so representable? What primes? What squares?

Examine the forms x^2+xy+y^2 and x^2-xy+y^2. These are essentially the same form, since we can obtain the second from the first by the transformation $x = X$ and $y = Y - X$ and the first from the second by the inverse of this substitution, namely $x = X$, $y = X + Y$. Thus, anything that we can say about the numbers representable by the first form can be said about the numbers representable by the second.

Study forms of the type $x^2 + by^2$, or more generally of the type $ax^2 + by^2$. Note that if $x = 2u$ and $y = 2v$ are even, then $x^2+xy+y^2 = (2u+v)^2+3v^2$, so we can partially relate the forms x^2+xy+y^2 and x^2+3y^2. More generally, if we start with $ax^2+hxy+by^2$ and make the substitution $X = px+qy$ and $Y = rx+sy$, we get the form $AX^2+HXY+BY^2$; under what circumstances is there an inverse substitution?

Exploration 4.6. Look at some examples of Pell's equation of the form $x^2 - dy^2 = 1$ and list the solution pairs (x_n, y_n) in nonnegative integers in increasing order of magnitude, with $(x_0, y_0) = (1, 0)$. Let $z_n = x_n y_n$. Look for patterns among the x_n, y_n, and z_n, perhaps using the case $d = 2$ as a model (see Exercise 1.2.4). In particular, check whether the sequences satisfy recursions of the form $t_{n+1} = at_n + bt_{n-1}$, and look for relationships among the pairs $(x_{n+1}+x_n, y_{n+1}-y_n)$ and $(x_{n+1}-x_n, y_{n+1}+y_n)$. Try to prove the generality of these patterns. Can you derive a process that will allow you to generate an unlimited number of solutions to the Pell's equation. Refer to Exploration 1.1 and compare the column headed p_nq_n with the solutions to $x^2 - 8y^2 = 1$. Can this be generalized?

Exploration 4.7. Let p be prime and $(x, y) = (u, v)$ be the fundamental solution of $x^2 - py^2 = 1$. Must p fail to divide v? This is an open question. The answer is *yes* for all primes not exceeding 6,000,000. If p is not prime, the answer is *no*. If you have access to a computer, try to find a counterexample.

4.6 Notes

1.10. See J.L. Lagrange, *Solutions d'un problème d'arithmétique*. Misc. Taurinensia 4 (1766-1769) = *Oeuvres* I, 671–731.

Section 4. A treatment of the bounds on the smallest solutions of $x^2 - dy^2 = k$ can be found in Mollin, pages 294–307.

4.7 Hints

1.2. Note that $ac - bd = a(c - d) + (a - b)d$; which factors are multiples of k?

1.3(d). Divide the integers in the solutions given in (c) by k.

1.4. Use the pigeonhole principle. Label k^2 "slots" by pairs (r, s), where $0 \leq r, s \leq k - 1$. Put the pair (x_i, y_i) in the slot (r, s) if and only if $x_i \equiv r$ and $y_i \equiv s \pmod{m}$.

1.6(a). Use the pigeonhole principle.

1.6(d). Use $|u^2 - dv^2| = |(u - v\sqrt{d})(u + v\sqrt{d})| < (1/v)(1/v + 2v\sqrt{d})$.

1.7. Use an induction argument. Suppose we have found already n distinct pairs. Since $|u_i - v_i\sqrt{d}|$ can never be zero, choose an integer N for which $1/N < |u_i - v_i\sqrt{d}|$ for $i = 1, \ldots, n$. Now apply Exercise 1.6.

1.8. Use the pigeonhole principle.

2.5. Follow the approach of Exercise 2.3.

3.13. Observe that $(p + q\sqrt{d})^3 = p(p^2 + 3q^2d) + q(3p^2 + q^2d)\sqrt{d}$. Consider the parenthetical terms modulo 8.

4.4. Use the fact that s_0 exceeds neither s_1 nor s_{-1}.

4.7(a). $d^2v^2s^2 = (dv^2)(ds^2) = (u^2 - 1)(r^2 - k)$.

4.7(b). Square $(u - 1)r \geq dvs$.

5

Tracking Down the Fundamental Solution

When d is a positive nonsquare integer, Section 4.1 assures us that there is a solution for $x^2 - dy^2 = 1$. However, we do not have much guidance on how actually to lay our hands on it. In this chapter we will develop algorithms that will lead to a solution.

If x and y are large positive integers for which $x^2 - dy^2 = 1$, then

$$\left|\frac{x}{y} - \sqrt{d}\right| = \frac{1}{y|x + y\sqrt{d}|}$$

will be pretty small. This means that x/y will be close to $\sqrt{d}$ so that we should be looking at close rational approximations to $\sqrt{d}$.

5.1 Adding Numerators and Denominators

In this section p, q, r, s will denote positive integers, and z a nonrational real number. We denote by $[a, b]$ the closed interval $\{t : a \leq t \leq b\}$.

Exercise 1.1. Suppose $\frac{p}{q} < \frac{r}{s}$. Explain why $\frac{p}{q} < \frac{p+r}{q+s} < \frac{r}{s}$.

Exercise 1.2. A weighted average of two real numbers u and v is a number of the form $(1 - t)u + tv$, where $0 \leq t \leq 1$.

(a) Verify that the usual average is obtained when $t = \frac{1}{2}$.

(b) Prove that a weighted average of u and v lies inside the interval (possibly at the endpoints) bounded by u and v.

(c) Let p, q, r, s be positive numbers. Express

$$\frac{p + r}{q + s}$$

as a weighted average of p/q and r/s.

Exercise 1.3. Suppose that $0 < \frac{p}{q} < z < \frac{r}{s}$.

(a) On a straight line segment of the real axis, illustrate possible positions of $\frac{p}{q}$, $\frac{r}{s}$, z, and $\frac{p+r}{q+s}$. Give numerical examples illustrating the possibilities.

(b) Is it always true that $|z - (p+r)/(q+s)|$ is smaller than at least one of $|z - p/q|$ and $|z - r/s|$? Explain.

(c) Observe that if z belongs to the interval $[p/q, r/s]$, then z belongs to one of the strictly smaller intervals $[p/q, (p+r)/(q+s)]$ and $[(p+r)/(q+s), r/s]$.

Exercise 1.4.

(a) Observe that $\frac{3}{1} < \pi < \frac{4}{1}$. Consider the intermediate fraction $\frac{3+4}{1+1} = \frac{7}{2}$. Which of the intervals $\left[3, \frac{7}{2}\right]$ and $\left[\frac{7}{2}, 4\right]$ contains π?

(b) The successive approximations for π will be the intermediate fractions formed by adding the numerators and denominators of under- and overestimates for $\pi = 3.1415926535\ldots$. Verify that when this is done, we obtain a sequence of approximations that begins 3/1, 4/1, 7/2, 10/3, 13/4, 16/5, 19/6, 22/7, 25/8, 47/15. Continue this sequence until you reach the term 1043/332.

(c) In this process, replacing a fraction by an equivalent fraction makes a difference to what follows. Verify that if in (b) we begin the sequence with 6/2 and 4/1, then the sequence begins with 6/2, 4/1, 10/3, 16/5, 22/7, 28/9, 50/16.

Exercise 1.5. Let us try to approximate $\sqrt{29}$ using the same process.

(a) Beginning with the under- and over-estimates 5/1 and 6/1 and continuing to produce new approximants by suitably adding numerators and denominators, obtain the sequence

$$\frac{5}{1}, \frac{6}{1}, \frac{11}{2}, \frac{16}{3}, \frac{27}{5}, \frac{43}{8}, \frac{70}{13}, \frac{97}{18}, \frac{167}{31}, \frac{237}{44}, \frac{307}{57}, \frac{377}{70}, \ldots.$$

(b) Let us assign to each of the fractions in the sequence one of the signs $-$ and $+$ according as it is less than or greater than $\sqrt{29}$. Verify that the assignment is

$$-, +, +, -, +, -, -, +, +, +, +, +$$

and that each fraction is obtained from the most recent $-$ fraction and the most recent $+$ fraction from among its predecessors.

(c) Verify that the positive rational x/y is less than $\sqrt{29}$ if and only if $x^2 - 29y^2$ is negative.

(d) Observe that we can rewrite the information given in (a) in a table

x	y	$x^2 - 29y^2$
5	1	−4
6	1	7
11	2	5
16	3	−5
27	5	4
43	8	−7
70	13	−1
97	18	13
167	31	20
237	44	25
307	57	28
377	70	29

(e) We can continue this table without making explicit reference to either $\sqrt{29}$ or fractions. The algorithm we are developing can be continued as follows: To find the next values of x and y, look at the last (x, y) in the list for which $x^2 - 29y^2 < 0$ and the last for which $x^2 - 29y^2 > 0$; add the corresponding values of x and the corresponding values of y together. Continue the list until you come to a solution in integers x and y of $x^2 - 29y^2 = 1$.

Exercise 1.6. Replace 29 in Exercise 1.4 by other values of d, in particular $d = 2, 3, 5, 6, 7, 8$. In each case, taking k to be the largest integer less than $\sqrt{d}$, continue the table

x	y	$x^2 - dy^2$
k	1	$k^2 - d \equiv a < 0$
$k + 1$	1	$(k+1)^2 - d \equiv b > 0$
$2k + 1$	2	$(2k+1)^2 - 4d = 2(a+b) - 1$

until you determine a solution of the equation $x^2 - dy^2 = 1$.

Exercise 1.7. It is possible to determine the list of values of $x^2 - dy^2$ in the table without having to compute the values of x and y each time. Using the notation of Exercise 1.5, determine first $a = k^2 - d$ and $b = (k+1)^2 - d$. The third entry is $2(a+b) - 1$.

Now we continue as follows:

Suppose we have obtained the value m in the list of values of $x^2 - dy^2$. Let h be the most recent value of the same sign as m, and p the most recent value of the opposite sign. We form a "difference" table:

$$\begin{array}{ccccc} h & & & & \\ & & m-h & & \\ m & & & & 2p \\ & & m-h+2p & & \\ 2m-h+2p & & & & \end{array}$$

The order of filling in the entries is $m, h, m-h, 2p, m-h+2p, 2m-h+2p$.

The first column starts off with h, m. The second column consists of differences of consecutive entries of the first column. The top element of the third column is $2p$; this column consists of differences of consecutive entries of the second column and its entries are set constantly to $2p$. Continue the table, working left from each occurrence of the constant second difference $2p$, until a change of sign occurs in the first column; the elements of the first column (including the first one with the new sign) constitute the extension of the sequence of values of $x^2 - dy^2$.

Here is what happens with $d = 29$. Start with $-4, 7, 5$. The first table is

$$\begin{array}{ccc} 7 & & \\ & -2 & \\ 5 & & -8 \\ & -10 & \\ -5 & & \end{array}$$

We have extended the sequence to $-4, 7, 5, -5$.

The second table is

$$\begin{array}{ccc} -4 & & \\ & -1 & \\ -5 & & 10 \\ & 9 & \\ 4 & & \end{array}$$

which gives the extention to $-4, 7, 5, -5, 4$.

The next few tables are

5		
	−1	
4		−10
	−11	
−7		

−5		
	−24	
−74		8
	6	
−1		8
	14	
13		

4		
	9	
13		−2
	7	
20		−2
	5	
25		−2
	3	
28		−2
	1	
29		−2
	−1	
28		−2
	−3	
25		−2
	−5	
20		−2
	−7	
13		−2
	−9	
4		−2
	−11	
−7		

Compare with the table in Exercise 1.5.

Exercise 1.8. Having obtained the third column in the table of Exercise 1.4(d), explain how to reconstruct the entries in the first two columns from the third entry on down.

Exercise 1.9. Verify that when $d = 54$, the sequence of values of $x^2 - 54y^2$ is

$$-5, 10, 9, -2, 19, 25, 27, 25, 19, 9, -5, 10, 1,$$
$$-18, -29, -38, -45, -50, -53, -54, -53, \ldots.$$

(Although the number 1 is our goal, the sequence can, of course, be continued further. You may wish to do this for a while to see what happens.) Use the information implicit in this sequence to construct a solution for $x^2 - 54y^2 = 1$.

Exercise 1.10. We can look at the generation of the sequence of values of $x^2 - dy^2$ in another way. Verify that the result of Exercise 3.1.2.(c) can be written

$$[(a+u)^2 - (b+v)^2 d] + [(a-u)^2 - (b-v)^2 d] = 2[(a^2 - b^2 d) + (u^2 - v^2 d)].$$

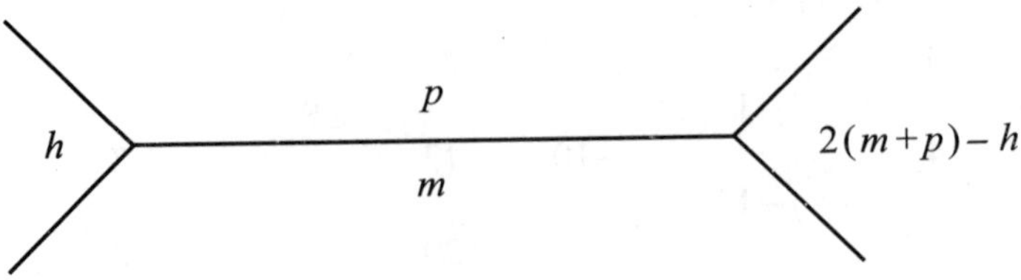

FIGURE 5.1.

Suppose, in the notation of Exercise 1.7, that $m = a^2 - b^2 d$ and $p = u^2 - v^2 d$.

(a) Explain why $h = (a-u)^2 - (b-v)^2 d$. (It might be helpful to refer to the table of Exercise 1.5(d).)

(b) Argue from the identity at the head of this exercise and Exercise 1.6 that the next entry after m is $(a+u)^2 - (b+v)^2 d = 2(m+p) - h$. This allows for a pictorial view. We can think of the values of $x^2 - dy^2$ as sitting in the cells surrounded by edges, where the values along a given edge are related to the values at the end of the edge by the following diagram:

For the case $d = 29$ we have the following diagram, where the values of $x^2 - dy^2$ are located on either side of a tree of edges separating positive and negative values: Make similar diagrams for other values of d.

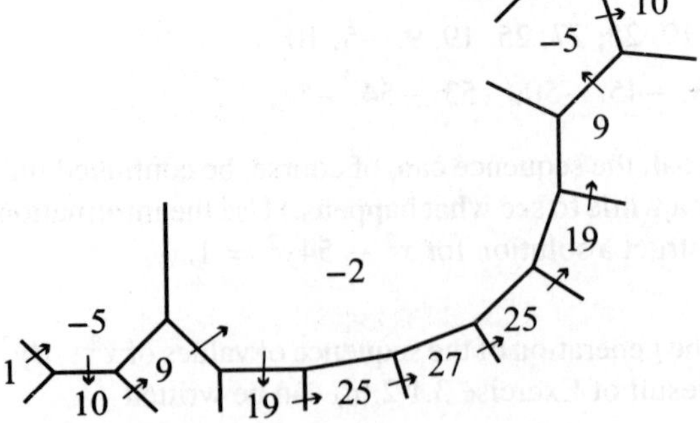

−5	10	9	−2	19
25	27	25	19	9
−5	10	1	−18	−29
−38	−45	−50	−53	

FIGURE 5.2.

5.2 Euclid's Algorithm

The algorithm of the last section can be accelerated. As a first step to this end, we study an algorithm introduced by Euclid in his *Elements* (VII:1,2; X:2). Suppose that z and w are two positive real numbers. We say that they are *commensurable* if both are positive integer multiples of the same number g; such a number g is said to be a *common divisor* of z and w. Not every pair of positive numbers is commensurable. Euclid's algorithm is a test to determine whether a given positive pair is commensurable and, if so, their greatest common divisor.

Exercise 2.1. Verify that 486 and 189 are commensurable and determine all of their common divisors. What is their greatest common divisor?

Exercise 2.2. Are the numbers 1 and $\sqrt{2}$ commensurable? (See Exercise 1.1.1.)

Exercise 2.3. A systematic way to find the greatest common divisor of two commensurable positive numbers z and w (not necessarily integers) with $z > w$ is to form a sequence of pairs (x, y) with $x > y \geq 0$ through this recursive process:

(i) the first pair is (z, w);

(ii) given a pair (x, y) in the sequence, the next pair is $(x - y, y)$ if $x - y \geq y$ and $(y, x - y)$ if $x - y < y$;

(iii) if a pair occurs in which the second entry is 0, then the process terminates.

We will show that the process terminates if and only if the pair (z, w) is commensurable, in which case the greatest common divisor is the first entry in the last pair (whose second entry is 0). The parts of this exercise will help you understand why this is so.

(a) Verify that the pair (486, 189) gives rise to the sequence (486, 189), (297, 189), (189, 108), (108, 81), (81, 27), (54, 27), (27, 27), (27, 0). Argue that any common divisor of the two numbers in any pair is a common divisor of the two numbers in the adjacent pair(s).

(b) For the pair (z, w), let (x, y) and (u, v) be consecutive pairs in its Euclidean sequence (generated using Euclid's algorithm). Show that any common divisor of x and y is a common divisor of u and v, and vice versa. Deduce that the greatest common divisor of x and y is equal to the greatest common divisor of u and v.

(c) Suppose that z and w are commensurable and that their greatest common divisor is d. Suppose that (x, y) and (u, v) are consecutive pairs in the Euclidean sequence with $x > y > 0$, $u > v > 0$. Prove that $u \leq x - d$. Deduce that if Euclid's algorithm is repeated often enough, we must arrive at a pair $(r, 0)$. Show that $r = d$.

(d) Suppose that the Euclidean sequence beginning with (z, w) terminates with the pair $(d, 0)$. Prove that d is the greatest common divisor of z and w.

Exercise 2.4. Use Euclid's algorithm to obtain the greatest common divisor of 27473 and 5627.

Exercise 2.5. We relate the problem of approximating the ratio z/w by common fractions a/b to Euclid's algorithm. The example we discuss is $z = 38$ and $w = 7$.

(a) Verify that the pairs obtained through Euclid's algorithm applied to 38 and 7 are $(38, 7)$, $(31, 7)$, $(24, 7)$, $(17, 7)$, $(10, 7)$, $(7, 3)$, $(4, 3)$, $(3, 1)$, $(2, 1)$, $(1, 1)$, $(1, 0)$.

(b) Consider the process described in Section 1 for approximating the fraction 38/7 by the fraction x/y. Begin with the pairs $(x, y) = (0, 1)$ and $(x, y) = (1, 0)$. Observe that $x/y < 38/7$ if and only if $7x - 38y < 0$. [Note that 1/0 is strictly speaking undefined; but for the sake of convention we can take it as larger than any real number and use it as a convenient starting upper bound for the number to be approximated.]

Verify that we can set out the numerators and denominators in the following table, analogous to that appearing in Exercise 5.1.5(d). In the fourth column we enter corresponding pairs from Euclid's algorithm.

x	y	$7x - 38y$	
0	1	-38	
1	0	7	
1	1	-31	$(38, 7)$
2	1	-24	$(31, 7)$
3	1	-17	$(24, 7)$
4	1	-10	$(17, 7)$
5	1	-3	$(10, 7)$
6	1	4	$(7, 3)$
11	2	1	$(4, 3)$
16	3	-2	$(3, 1)$
27	5	-1	$(2, 1)$
38	7	0	$(1, 1)$

(c) Examine the data in parts (a) and (b). In (a), note how many consecutive pairs there are whose second entries are the same and compare this with the number of consecutive terms in the third column of the table in (b) of the same sign. How do you account for your observation?

Exercise 2.6. Examine the general situation in which we require only that z and w be positive real numbers (not necessarily integers or even rationals) with $(r + 1)w > z > rw$ for some positive integer r.

(a) Verify that the table of numerators and denominators of fractions approximating z/w begins as follows; in the fourth column we list the pairs arising in Euclid's algorithm for the greatest common divisor of z and w.

x	y	$wx - zy$	Euclid's algorithm
0	1	$-z < 0$	
1	0	$w > 0$	
1	1	$w - z < 0$	(z, w)
2	1	$2w - z < 0$	$(z - w, w)$
3	1	$3w - z < 0$	$(z - 2w, w)$
...	...	...	...
r	1	$rw - z < 0$	$(z - (r - 1)w, w)$
$r + 1$	1	$(r + 1)w - z > 0$	$(w, z - rw)$
$2r + 1$	2	$(2r + 1)w - 2z$	...

Observe that the number of pairs with second entry w is equal to the number of consecutive negative terms in the third column. Compare the difference in the entries of the pairs in the fourth column to the values in the third column.

(b) Let us imagine that the table of part (a) is continued to some point at which a sign change again occurs in the value of $wx - zy$. As an induction hypothesis, we suppose that the entries appear in the form below; with no loss of generality, suppose that the sign in the third column changes from negative to positive.

x	y	$wx - zy$	Euclid's algorithm
a	b	$aw - by = u - t < 0$	(t, u)
c	d	$cw - dz = u - v > 0$	$(u, v) = (u, t - u)$

Suppose that $(s + 1)v > u > sv$ for some positive integer s. Argue that the table continues, yielding s consecutive positive terms in the third column.

x	y	$wx - zy$	Euclid's algorithm
a	b	$aw - by = u - t < 0$	(t, u)
c	d	$cw - dz = u - v > 0$	$(u, v) = (u, t - u)$
$a + c$	$b + d$	$2u - (t + v) = u - 2v > 0$	$(u - v, v)$
$2a + c$	$2b + d$	$3u - (2t + v) = u - 3v > 0$	$(u - 2v, v)$
$3a + c$	$3b + d$	$4u - (3t + v) = u - 4v > 0$	$(u - 3v, v)$
...	...	...	...
$(s - 1)a + c$	$(s - 1)b + d$	$u - sv > 0$	$(u - (s - 1)v, v)$
$sa + c$	$sb + d$	$u - (s + 1)v < 0$	$(v, u - sv)$

(If the sign change at the top is from positive to negative, the table is easily adapted. If $u = sv$, the table terminates with a 0 in the third column.)

(c) Observe that each block of consecutive terms of the same sign in the third column corresponds to a block of consecutive terms in Euclid's algorithm for which the second entry remains the same.

(d) Fill in the fourth column of the table in Exercise 2.4.

Exercise 2.7.

(a) Euclid's algorithm proceeds by a succession of subtractions. We can streamline the method by using division in place of repeated subtractions of the same quantity. Suppose that

$$z = a_0 w + w_1,$$

where $a_0 w \le z < (a_0 + 1)w$, $0 \le w_1 < w$. Then $a_0 = \lfloor z/w \rfloor$. Then Euclid's algorithm gives rise to the pairs (z, w), $(z - w, w)$, ..., $(z - (a_0 - 1)w, w)$, (w, w_1). We can proceed with

$$w = a_1 w_1 + w_2,$$

where $a_1 w_1 \le w < (a_1 + 1)w_1, 0 \le w_2 < w_1$ (so that $a_1 = \lfloor w/w_1 \rfloor$). This relation encapsulates the next block of a_1 pairs whose second entries are w_1.

Argue that Euclid's algorithm can be abbreviated to a succession of equations and corresponding pairs

$$\begin{array}{ll} z = a_0 w + w_1 & (z, w) \\ w = a_1 w_1 + w_2 & (w, w_1) \\ w_1 = a_2 w_2 + w_3 & (w_1, w_2) \\ w_2 = a_3 w_3 + w_4 & (w_2, w_3) \\ \cdots & \cdots \end{array}$$

where $z > w > w_1 > w_2 > \cdots$ and $a_i = \lfloor w_{i-1}/w_i \rfloor$ for each positive i $(w_0 \equiv w)$.

(b) Prove that z and w are commensurable if and only if some w_{k+1} is zero, in which case the process terminates with an equation $w_{k-1} = a_k w_k$. Argue that in this case the greatest common divisor of z and w is w_k.

(c) Prove that the positive reals z and w are not commensurable if and only if z/w is irrational if and only if the process fails to terminate with a zero remainder.

(d) Apply the abbreviated algorithm to finding the greatest common divisor of 38 and 7, and of 27473 and 5627.

5.3 Continued Fractions

This section pursues the work of the last in the special case that $w = 1$. Suppose that z is a positive irrational and that a_0 is the largest nonnegative integer that does not exceed z. We can write $z = a_0 + w_1$, where w_1 is positive and less than 1. This decomposition will be the seed that generates the algorithm discussed in this section.

Exercise 3.1.

(a) Show that $z = a_0 + 1/u_1$ where $u_1 > 1$.

(b) Repeat the process for u_1 to obtain a representation of the form

$$u_1 = a_1 + \frac{1}{u_2},$$

where $u_2 > 1$. Deduce that

$$z = a_0 + \cfrac{1}{a_1 + \frac{1}{u_2}}$$

or, for short,

$$z = a_0 + 1/a_1 + 1/u_2.$$

(c) Verify that we can repeat the process repeatedly to obtain

$$\begin{aligned} z &= a_0 + 1/a_1 + 1/a_2 + 1/u_3 = a_0 + 1/a_1 + 1/a_2 + 1/a_3 + 1/u_4 = \cdots \\ &= a_0 + 1/a_1 + 1/a_2 + 1/a_3 + \cdots + 1/a_n + 1/u_{n+1}, \end{aligned}$$

where $a_1, a_2, \ldots, a_n$ are positive integers and $u_3, \ldots, u_{n+1}$ are real numbers exceeding 1. We adopt the convention that each slash is the bar of a fraction whose denominator is everything that follows it.

Exercise 3.2. Verify that the process of Exercise 3.1 applied to π yields $\pi = 3 + 1/7 + 1/15 + 1/1 + 1/292 + 1/1 + 1/1 + 1/1 + 1/2 + \cdots$.

This can be done conveniently on a pocket calculator or computer by entering the number π (i.e., a sufficiently close decimal approximation) and repeating the following loop:

(i) record the integer part of the number;
(ii) subtract the integer part of the number;
(iii) invert (press the $1/x$ button);
(iv) return to (i).

Exercise 3.3.

(a) Using a pocket calculator, apply the process of Exercise 3.1 to the number $\sqrt{29}$.

(b) An alternative way of doing part (a) is available to us because of the form of $\sqrt{29}$ as a surd. Verify the following:

$$\sqrt{29} = 5 + \left(\sqrt{29} - 5\right) = 5 + \frac{4}{\sqrt{29} + 5},$$

$$\frac{\sqrt{29} + 5}{4} = 2 + \frac{\sqrt{29} - 3}{4} = 2 + \frac{5}{\sqrt{29} + 3},$$

$$\frac{\sqrt{29} + 3}{5} = 1 + \frac{\sqrt{29} - 2}{5} = 1 + \frac{5}{\sqrt{29} + 2},$$

$$\frac{\sqrt{29} + 2}{5} = 1 + \frac{\sqrt{29} - 3}{5} = 1 + \frac{4}{\sqrt{29} + 3},$$

$$\frac{\sqrt{29} + 3}{4} = 2 + \frac{\sqrt{29} - 5}{4} = 2 + \frac{1}{\sqrt{29} + 5},$$

$$\sqrt{29} + 5 = 10 + \left(\sqrt{29} - 5\right) = 10 + \frac{4}{\sqrt{29} + 5},$$

and use these computations to check the result of (a).

Exercise 3.4. From the previous exercises, we see that $\sqrt{29}$ can be written in continued fraction form as $5+1/\underline{2}+1/\underline{1}+1/\underline{1}+1/\underline{2}+1/\underline{10}+1/\underline{2}+1/\underline{1}+1/\cdots$ where the underlined numbers cycle through 2, 1, 1, 2, 10. We form successive approximations (called *convergents* to $\sqrt{29}$) by stopping the representation at each of the underlined numbers, thus:

$$5,\ 5+1/2,\ 5+1/2+1/1,\ 5+1/2+1/1+1/1,\ 5+1/2+1/1+1/1+1/2,\ldots.$$

(a) Verify that the sequence of convergents is

$$\frac{5}{1},\ \frac{11}{2},\ \frac{16}{3},\ \frac{27}{5},\ \frac{70}{13},\ \frac{727}{135},\ \frac{1524}{283},\ \frac{2251}{418},\ \frac{3775}{701},\ \frac{9801}{1820},\ldots.$$

(b) Explain why these convergents are alternately less than and greater than $\sqrt{29}$.

Exercise 3.5. For irrational z we can carry out the same procedure to get $z = a_0 + 1/a_1 + 1/a_2 + 1/a_3 + 1/a_4 + \cdots + 1/a_n + \cdots$. Define p_n and q_n to be the numerator and denominator of the nth convergent when written in lowest terms:

$$\frac{p_n}{q_n} = a_0 + 1/a_1 + 1/a_2 + 1/a_3 + \cdots + 1/a_n \quad (n = 0, 1, 2, \ldots).$$

Verify that

$$\begin{aligned} p_0 &= a_0, & q_0 &= 1,\\ p_1 &= a_0a_1 + 1, & q_1 &= a_1,\\ p_2 &= a_0a_1a_2 + a_0 + a_2, & q_2 &= a_1a_2 + 1,\\ p_3 &= a_0a_1a_2a_3 + a_0a_3 + a_2a_3 + a_0a_1 + 1, & q_3 &= a_1a_2a_3 + a_1 + a_3. \end{aligned}$$

Be sure to check that the greatest common divisor of p_i and q_i is 1 in each case.

Exercise 3.6. There is a convenient algorithm for obtaining each p_n and each q_n from its predecessor, once p_0, q_0, p_1, and q_1 are known. For $n \geq 2$, it turns out that

$$p_n = a_n p_{n-1} + p_{n-2}, \quad q_n = a_n q_{n-1} + q_{n-2}.$$

Let us understand why this is so.

Suppose we define $\phi_n(a_0, a_1, \ldots, a_n)$ for $n = -1, 0, 1, 2, \ldots$ by the following:

$$\phi_{-1} = 1, \quad \phi_0(a_0) = a_0, \quad \phi_1(a_0, a_1) = a_0a_1 + 1,$$

$$\phi_n(a_0, a_1, a_2, \ldots, a_n) = a_n\phi_{n-1}(a_0, a_1, \ldots, a_{n-1}) + \phi_{n-2}(a_0, a_1, \ldots, a_{n-2}).$$

Observe that when $n = 0, 1, 2, 3$, then $p_n = \phi_n(a_0, a_1, a_2, \ldots, a_n)$ and $q_n = \phi_{n-1}(a_1, a_2, \ldots, a_n)$.

(a) Suppose that

$$\frac{p_n}{q_n} = \frac{\phi_n(a_0, a_1, \ldots, a_n)}{\phi_{n-1}(a_1, \ldots, a_n)}$$

has been established for $n = 1, 2, 3, \ldots, m$, where m is a positive integer.

By considering $a_m + \frac{1}{a_{m+1}}$ as a single entity, verify that

$$\begin{aligned}
\frac{p_{m+1}}{q_{m+1}} &= \frac{\phi_m(a_0, a_1, \ldots, a_m + 1/a_{m+1})}{\phi_{m-1}(a_1, \ldots, a_m + 1/a_{m+1})} \\
&= \frac{(a_m + 1/a_{m+1})\phi_{m-1}(a_0, a_1, \ldots, a_{m-1}) + \phi_{m-2}(a_0, a_1, \ldots, a_{m-2})}{(a_m + 1/a_{m+1})\phi_{m-2}(a_1, \ldots, a_{m-1}) + \phi_{m-3}(a_1, \ldots, a_{m-2})} \\
&= \frac{a_{m+1}\phi_m(a_0, a_1, \ldots, a_m) + \phi_{m-1}(a_0, a_1, \ldots, a_{m-1})}{a_{m+1}\phi_{m-1}(a_1, a_2, \ldots, \phi_m) + \phi_{m-2}(a_1, a_2, \ldots, a_{m-1})} \\
&= \frac{\phi_{m+1}(a_0, a_1, \ldots, a_{m+1})}{\phi_m(a_1, \ldots, a_{m+1})}.
\end{aligned}$$

(b) Prove that $\phi_n(a_0, a_1, \ldots, \phi_n) = \phi_n(a_n, a_{n-1}, \ldots, \phi_0)$ and deduce that

$$\phi_n(a_0, a_1, \ldots, a_n) = a_0\phi_{n-1}(a_1, a_2, \ldots, a_n) + \phi_{n-2}(a_2, a_3, \ldots, a_n).$$

(c) Prove by induction that for any choice of positive integers $a_0, a_1, \ldots$, the greatest common divisor of any two consecutive terms in the sequence $\{\phi_n(a_0, a_1, \ldots, a_n) : n = 0, 1, 2, \ldots\}$ is 1.

(d) Prove, for $n = 1, 2, \ldots$, that $p_n = \phi_n(a_0, \ldots, a_n)$ and $q_n = \phi_{n-1}(a_1, \ldots, a_n)$.

(e) Prove the recursion formula for p_n and q_n given at the beginning of the exercise.

Exercise 3.7.

(a) Prove by induction that for each positive integer n,

$$p_{n+1}q_n - p_nq_{n+1} = (-1)^n.$$

(b) Deduce from (a) that

$$\lim_{n\to\infty}\left[\frac{p_{n+1}}{q_{n+1}} - \frac{p_n}{q_n}\right] = 0.$$

(c) Prove that the convergents p_n/q_n are alternately less than and greater than z.

(d) Deduce that

$$\left|z - \frac{p_n}{q_n}\right| \leq \left|\frac{p_{n+1}}{q_{n+1}} - \frac{p_n}{q_n}\right| < \frac{1}{q_n^2},$$

so that

$$z = \lim_{n\to\infty}\frac{p_n}{q_n}.$$

(e) Prove that

$$u_n = \lim_{k\to\infty}\frac{\phi_k(a_n, a_{n+1}, \ldots, a_{n+k})}{\phi_{k-1}(a_{n+1}, a_{n+2}, \ldots, a_{n+k})}.$$

Exercise 3.8. Using the formula given in Exercise 3.6 and $d = 29$, verify the following table:

a_n	a_n	p_n	q_n	$p_n^2 - 29q_n^2$
0	5	5	1	−4
1	2	11	2	5
2	1	16	3	−5
3	1	27	5	4
4	2	70	13	−1
5	10	727	135	

Continue the table until the number 1 appears in the final column.

Exercise 3.9. Compare the table in Exercise 3.8 with the table in Exercise 1.5(d). What is the significance of the numbers a_n with respect to the earlier table? Review Exercise 2.5 to understand the role of a_n.

Exercise 3.10. Use the procedure of Exercise 3.8 to determine the fundamental solution of $x^2 - dy^2 = 1$ for other values of d.

Exercise 3.11. Suppose that $z = a_0 + 1/a_1 + 1/a_2 + \cdots + 1/a_n + \cdots$ and $w = b_0 + 1/b_1 + 1/b_2 + \cdots + 1/b_n + \cdots$, where either development could be finite. Let the convergents for z be p_n/q_n.

(a) Prove that if

$$\frac{p_0}{q_0} \leq w < \frac{p_1}{q_1},$$

then $b_0 = a_0$.

(b) Prove that if

$$\frac{p_2}{q_2} \leq w < \frac{p_1}{q_1},$$

then $b_0 = a_0$ and $b_1 = a_1$.

(c) Prove, by induction on m, that if w lies between p_m/q_m and p_{m+1}/q_{m+1}, then $b_0 = a_0, b_1 = a_1, \ldots, b_m = a_m$, so that the first $m + 1$ convergents for w are p_i/q_i, where $i = 0, 1, \ldots, m$. ♠

In the following exercises we show that the convergents of the continued fraction for a irrational z are best approximations for the size of the denominators. In the following, $\{p_n/q_n\}$ is the sequence of convergents for the irrational number z.

Exercise 3.12. Suppose that a and b are integers with $b > 0$ and $|zb - a| < |zq_n - p_n|$ for some positive integer n. We wish to show that $b \geq q_{n+1}$. The argument will be by contradiction, so we assume the contrary, i.e., that $b < q_{n+1}$.

(a) Prove that the system

$$q_n x + q_{n+1} y = b,$$
$$p_n x + p_{n+1} y = a,$$

has a solution (x, y) in integers.

(b) Prove that this solution cannot have $x = 0$.

(c) Verify that

$$|zb - a| = |x(q_n z - p_n) + y(q_{n+1} z - p_{n+1})|$$

and argue that $y \neq 0$ for any solution of (a).

(d) Prove that $xy < 0$ for any solution of (a).

(e) From (d), note that

$$|zb - a| = |x(q_n z - p_n)| + |y(q_{n+1} z - p_{n+1}| \geq |x||q_n z - p_n|$$

and obtain a contradiction to our hypothesis. Deduce that $b < q_{n+1}$.

Exercise 3.13. Prove that if a and b are integers with $b > 0$, and $|z - a/b| < |z - p_n/q_n|$ for some $n \geq 1$, then $b > q_n$.

Exploration 5.1. The following is an algorithm due to Tenner for obtaining the values $\{a_n\}$ involved in the continued fraction representation for $\sqrt{d}$. Suppose that $k = \lfloor \sqrt{d} \rfloor$ and $d = k^2 + r$. We form a table with six columns:

I	II	III		IV	V	VI
k	$\times$	k	$=$	d	$-$	r
		$\dots$				
a	b	c		u	v	w
A	B	C		U	V	W

where a given row (uppercase letters) is obtained from the previous row (lowercase letters) by the following relations:

$$k + c = Aw + B, \quad 0 \leq B < w,$$
$$C = k - B,$$
$$U = C^2,$$
$$V = d - U,$$
$$W = V/w.$$

Column I gives the successive values of a_n.

Verify that the table for $d = 29$ is the following:

I	II	III		IV	V	VI
5	×	5	=	29	—	4
2	2	3		9	20	5
1	3	2		4	25	5
1	2	3		9	20	4
2	0	5		25	4	1
10	0	5		25	4	4
2	2	3		9	20	5
		...				

Try this method for other values of d. Why does it work? What is the significance of the numbers in Column VI?

5.4 Periodic Continued Fractions

Does the continued fraction expansion of $\sqrt{d}$ always deliver a fundamental solution of Pell's equation? It turns out that those z that are roots of quadratic equations with integer coefficients are characterized by sequences $\{a_n\}$ that are eventually periodic. In this section the symbols z, u_i, a_i have the meanings assigned to them in Section 5.3.

Exercise 4.1. Suppose that in the continued fraction expansion of z we obtain a sequence $\{a_n\}$ of integers that is periodic from the term a_r on; i.e., for some positive integer h, $a_{n+h} = a_n$ for $n \geq r$.

(a) Prove that $u_n = u_{n+h}$ for $n \geq r$.

(b) Using $z = a_0 + 1/a_1 + 1/a_2 + \cdots + 1/a_{n-1} + 1/u_n$ and Exercise 3.6, prove that

$$z = \frac{u_r p_{r-1} + p_{r-2}}{u_r q_{r-1} + q_{r-2}} = \frac{u_r p_{r+h-1} + p_{r+h-2}}{u_r q_{r+h-1} + q_{r+h-2}}.$$

(c) Show that there are integers A, B, C with $A \neq 0$ for which u_r is a root of the quadratic equation $At^2 + Bt + C = 0$, so that $u_r = p \pm \sqrt{q}$ for some rationals p and q. Deduce that z has the same form.

(d) Verify that a number z has the form $p \pm \sqrt{q}$ for some rationals p and q if and only if z is a root of a quadratic equation with integer coefficients.

Exercise 4.2.

(a) Suppose that z is irrational and that $Az^2 + Bz + C = 0$ for some integers A, B, C. Prove that there are integers A_n, B_n, C_n such that

$$A_n u_n^2 + B_n u_n + C_n = 0.$$

Indeed, verify that

$$\begin{aligned} A_n &= Ap_{n-1}^2 + Bp_{n-1}q_{n-1} + Cq_{n-1}^2, \\ B_n &= 2Ap_{n-1}p_{n-2} + B(p_{n-1}q_{n-2} + p_{n-2}q_{n-1}) + 2Cq_{n-1}q_{n-2}, \\ C_n &= Ap_{n-2}^2 + Bp_{n-2}q_{n-2} + Cq_{n-2}^2. \end{aligned}$$

(b) Why is $A_n \neq 0$?

(c) Verify that $B_n^2 - 4A_nC_n = B^2 - 4AC$.

(d) The next step is to deduce that the numbers A_n, B_n, C_n have an upper bound independent of n, so that there are finitely many possibilities for the triple (A_n, B_n, C_n). Recall from Exercise 3.7 that

$$\left| z - \frac{p_n}{q_n} \right| < \frac{1}{q_n^2},$$

so that

$$-\frac{1}{q_n} < p_n - zq_n < \frac{1}{q_n}$$

for all n. Use this to deduce that

$$\begin{aligned} |A_n| &< 2|Az| + |A| + |B|, \\ |C_n| = |A_{n+1}| &< 2|Az| + |A| + |B|, \\ |B_n|^2 &< 4[2|Az| + |A| + |B|]^2 + |B^2 - 4AC|. \end{aligned}$$

(e) Use the pigeonhole principle to deduce that the triple (A_n, B_n, C_n) can assume at most finitely many values, so that for some positive integers r and h, we must have that

$$(A_r, B_r, C_r) = (A_{r+h}, B_{r+h}, C_{r+h}).$$

Deduce that $\{u_n\}$ and therefore that $\{a_n\}$ is periodic with period h from some point on. ♠

The following exercises investigate the same result another way. We will focus on using surd manipulations in obtaining the continued fraction expansion. To get a feel for the result of the next exercise, review the data given in Exercise 3.3 for the continued fraction expansion of $\sqrt{29}$. When the numerators are rationalized, observe that a fraction is obtained whose numerator is an integer. For example, $(\sqrt{29} - 3)/4$ is such that 4 divides $29 - 3^2$, and so $(\sqrt{29} - 3)/4$ equals a fraction with denominator $\sqrt{29} + 3$ whose numerator is the integer 5.

Exercise 4.3. Suppose that z is a positive irrational real solution of a quadratic equation with integer coefficients.

(a) Prove that z has the form $(k_0 + \sqrt{d})/h_0$, where d is a positive integer, h_0 and k_0 are integers, and $d - k_0^2$ is a multiple of h_0. Give this form for both of the roots of the quadratic equation $7x^2 - 8x + 2 = 0$.

(b) Suppose that $z = a_0 + (1/u_1)$ with a_0 an integer and $u_1 > 1$. Verify that

$$u_1 = \frac{h_0[-(k_0 - a_0 h_0) + \sqrt{d}]}{d - (k_0 - a_0 h_0)^2}.$$

Express this in the form $(k_1 + \sqrt{d})/h_1$, where h_1 and k_1 are integers with $h_1 \neq 0$ and $d - k_1^2$ is a multiple of h_1.

(c) Continue in this fashion to show that each u_i arising in the continued fraction expansion of z has the form

$$\frac{k_i + \sqrt{d}}{h_i},$$

where h_i and k_i are integers, $h_i \neq 0$, and $h_i h_{i-1} = d - k_i^2$.

(d) Illustrate (c) for each of the roots of the equation $7x^2 - 8x + 2 = 0$.

Exercise 4.4. Following the notation of Exercise 4.3, let $v_i = (k_i - \sqrt{d})/h_i$ be the "surd conjugate" of u_i and w be the surd conjugate of z.

(a) Explain why

$$w = \frac{v_n p_{n-1} + p_{n-2}}{v_n q_{n-1} + q_{n-2}}.$$

(b) Verify that

$$v_n = -\frac{q_{n-2}}{q_{n-1}}\left(\frac{w - (p_{n-2}/q_{n-2})}{w - (p_{n-1}/q_{n-1})}\right)$$

and argue that v_n is negative when n exceeds some positive integer N.

(c) Show that $u_n - v_n$ is positive for $n > N$, so that h_n is also positive for these values of n.

(d) Use the fact that $h_n h_{n-1} = d - k_n^2$ to deduce that the pair (h_n, k_n) can assume at most finitely many distinct values.

(e) Show that there are positive integers r and s for which $s > r$ and $(h_r, k_r) = (h_s, k_s)$. Deduce that $u_r = u_s$, so that for $n \geq r$, $a_{n+s-r} = a_n$ (i.e., the sequence $\{a_n\}$ is eventually periodic).

Exercise 4.5. This exercise will investigate when the continued fraction expansion of the positive irrational z is purely periodic, so that for some positive integer s,

$$z = a_0 + 1/a_1 + 1/a_2 + \cdots + 1/a_{s-1} + 1/a_0 + 1/a_1 + \cdots.$$

This occurs when

$$z = a_0 + 1/a_1 + 1/a_2 + \cdots + 1/a_{s-1} + 1/z,$$

or when

$$z = \frac{z p_{s-1} + p_{s-2}}{z q_{s-1} + q_{s-2}}.$$

Determine a quadratic equation $f(x) = 0$ with integer coefficients for which z and w are the roots. By looking at $f(0)$ and $f(-1)$, deduce that $-1 < w < 0$. Explain why a_0 must be positive. Deduce that $z > 1$.

Exercise 4.6. The converse of the result in Exercise 4.5 also holds. Suppose that z and its surd conjugate w are the roots of a quadratic equation with integer coefficients and that

$$-1 < w < 0 < 1 < z.$$

(a) Explain why $1/v_{i+1} = v_i - a_i$ for each positive integer i and prove by induction that $-1 < v_i < 0$. Deduce that $a_i = \lfloor -1/v_{i+1} \rfloor$. (Note that this relates a_i to v_{i+1}, which has a greater index.)
(b) By Exercises 4.2 and 4.3, since z is a root of a quadratic equation with integer coefficients, there exist integers r and s for which $u_r = u_s$ with $r < s$. Deduce from $v_r = v_s$ that $a_{r-1} = a_{s-1}$, so that $u_{r-1} = u_{s-1}$.
(c) Prove that $z = u_{s-r}$, so that $\{a_n\}$ is periodic with the initial segment $\{a_0, a_1, \ldots, a_{s-r-1}\}$ repeated.
(d) Let d be a positive nonsquare integer. Prove that $\sqrt{d} + \lfloor \sqrt{d} \rfloor$ has a purely periodic continued fraction expansion of the form

$$\sqrt{d}+\lfloor\sqrt{d}\rfloor = 2a_0+1/a_1+1/a_2+1/a_3+\cdots+1/a_{s-1}+1/2a_0+1+a_1/1+\cdots$$

with $a_0 = \lfloor \sqrt{d} \rfloor$. What is the continued fraction expansion of $\sqrt{d}$?

Exercise 4.7. In Exercise 4.3, suppose that $z = \sqrt{d}$, so that $k_0 = 0$ and $h_0 = 1$.
(a) Prove that for $n \geq 0$,

$$\sqrt{d} = \frac{(k_{n+1} + \sqrt{d})p_n + h_{n+1}p_{n-1}}{(k_{n+1} + \sqrt{d})q_n + h_{n+1}q_{n-1}}.$$

(b) Manipulate the equation in (a) into the form $A + B\sqrt{d} = 0$, where A and B are integers. Why should A and B vanish? Deduce that

$$p_n^2 - dq_n^2 = (-1)^{n-1}h_{n+1}.$$

(c) If the continued fraction expansion of $\sqrt{d}$ has ultimate period s, prove that for each positive integer m,

$$p_{ns-1}^2 - dq_{ns-1}^2 = (-1)^{ns}.♠$$

We have thus established that $x^2 - dy^2 = 1$ is solvable whenever d is a positive nonsquare integer and shown how the continued fraction method provides an algorithm for obtaining it. You might try this to obtain solutions for $x^2 - dy^2 = 1$ for values of d that otherwise have presented difficulties.

Exploration 5.2. Factoring integers. Suppose that we wish to factor the integer 1037. If we can find p and q for which $p^2 - 1037q^2 = r^2$, a perfect square, then $1037q^2 = p^2 - r^2 = (p-r)(p+r)$, and it may be possible to find factors of 1037 distributed among those of $p-r$ and $p+r$. The Euclidean algorithm can be used to find the greatest common divisor of the pairs $(1037, p-r)$ and $(1037, p+r)$. Thus, we can search among the values of $p_n^2 - 1037q_n^2$ (or equivalently, by Exercise 4.7(b), among the values of $(-1)^{n-1}h_{n+1}$) for perfect squares. Use the algorithm to obtain the possibility $(p, q, r) = (129, 4, 7)$; observe that $1037 \times 16 = 129^2 - 7^2 = 122 \times 136$. Now find the greatest common divisor of the pairs (1037, 122) and (1037, 136). Try this out to factor other large integers d. If you are unsuccessful, consider replacing d by a multiple md with m and d coprime.

5.5 The Polynomial Case

We can look at an equation $x^2 - d(t)y^2 = 1$ in two ways: either as a parameterized numerical equation or as a polynomial equation. (Go back to Exploration 2.7 for some examples.) In the former case we can let t have numerical values and the look for patterns in the solutions that can then be expressed in polynomial form.

We can look at polynomials as more than carriers of numerical patterns. They are mathematical entities in their own right. Polynomials can be added, subtracted, multiplied, and divided in the way that integers can, and we can consider $x^2 - d(t)y^2 = 1$ as an equation in which polynomial solutions are to be found. Not every possibility for $d(t)$ will admit nontrivial solutions; can you think of any? The polynomial version of Pell's equation does not seem to be well treated in the literature, and in this section we will study a few examples. We can try to imitate the continued fraction approach, but this is hampered by the lack of a magnitude for polynomials, which allows us to use the floor function for numbers. We can try comparing possible developments with those obtained when numbers are substituted for the variable. A final approach is to use the method of undetermined coefficients; assume that you have a Pell's equation with polynomials of various degrees, and then equate coefficients of various powers of the variable t that arise in the expansion. We do not pursue this in the exercises, but you might want to look at some examples, particularly if you have computer software capable of dealing with algebraic expressions.

Exercise 5.1. Beginning with the observation that

$$\sqrt{t^2+1} = t + \left(\sqrt{t^2+1} - t\right) = t + \frac{1}{\sqrt{t^2+1}+t},$$

obtain the continued fraction development

$$\sqrt{t^2+1} = t + 1/2t + 1/2t + 1/2t + \cdots.$$

By examining its convergents, obtain polynomial solutions of

$$x^2 - (t^2+1)y^2 = \pm 1.$$

Exercise 5.2.

(a) Imitate the process of Exercise 5.1 for $\sqrt{t^2+c}$ to generate polynomial solutions for $x^2-(t^2+c)y^2=1$. Compare your solution with what you found in Exercise 5.1.

(b) Do (a) in the special case $c=2$.

Exercise 5.3. Beginning with the observation that

$$\sqrt{t^2-1}=(t-1)+\left[\sqrt{t^2-1}-(t-1)\right]=(t-1)+\frac{2(t-1)}{\sqrt{t^2-1}+(t-1)},$$

obtain the continued fraction development

$$\sqrt{t^2-1}=(t-1)+1/1+1/2(t-1)+1/1+1/2(t-1)+1/1+\cdots.$$

Use this to obtain polynomial solutions to $x^2-(t^2-1)y^2=1$.

Exercise 5.4. Consider the equation $x^2-(4t^2+12t+5)y^2=1$.

(a) Verify, for $t>0$, that

$$2t+2<\sqrt{4t^2+12t+5}<2t+3.$$

(b) Use the observation in (a) to imitate the continued fraction algorithm and obtain a solution for the equation.

Exercise 5.5.

(a) Let $p(t)=m^2t^4+2mnt^3+n^2t^2+mt+n=(mt^2+nt)^2+(mt+n)$ where $mn\neq 0$. Obtain the continued fraction $\sqrt{p(t)}=(mt^2+nt)+1/2t+1/2(mt^2+nt)+1/2t+\cdots$ and the solution

$$(x,y)=(2mt^3+2nt^2+1,2t)$$

for $x^2-p(t)y^2=1$.

(b) Find a solution for

$$x^2-(m^2t^4+2mnt^3+n^2t^2+2mt+2n)y^2=1.$$

Exercise 5.6.

(a) Solve $x^2-(4t^2+4t+5)y^2=-1$.

(b) Solve $x^2-(4t^4+4t+2)y^2=-1$.

(c) Solve $x^2-(16t^4+8t^3+9t^2+6t+2)y^2=-1$. ♠

Next we look at the case $d(t)=t^2+3$. The continued fraction approach begins with

$$\sqrt{t^2+3}=t+\frac{3}{\sqrt{t^2+3}+t},$$

so we need to deal with $(\sqrt{t^2+3}+t)/3$ at the next stage. This is about equal to $2t/3$, not always an integer when t is an integer. In order to clarify the situation, we subdivide the problem into three cases corresponding to the remainder that a numerical value of t might have upon division by 3: $t = 3s$, $t = 3s + 1$, and $t = 3s + 2$. We then make a similar analysis when $d(t) = t^2 - 4$.

Exercise 5.7.
(a) Verify that the continued fraction expansion for $\sqrt{9s^2 + 3}$ is

$$3s + 1/2s + 1/6s + 1/2s + 1/6s + \cdots$$

and thus obtain solutions for $x^2 - (9s^2 + 3)y^2 = 1$.
(b) Generalize (a) to the situation $d = t^2 + k$ with $t = ks$.
(c) For which numerical values of d can we obtain solutions to $x^2 - dy^2 = 1$ by making substitutions for s and k in (a) and (b)?

Exercise 5.8. When $t = 3s + 2$, then $t^2 + 3 = 9s^2 + 12s + 7$.
(a) Prepare a table giving the values of a_n when $\sqrt{3s + 2}$ has the form $a_0 + 1/a_1 + 1/a_2 + 1/a_3 + \cdots$ for $s = 0, 1, 2, 3, 4, 5, \ldots$. (See Exploration 5.1 at the end of Section 3 for a convenient algorithm.)
(b) Verify that

$$\sqrt{9s^2 + 12s + 7} = (3s + 2) + 1/(2s + 1) + 1/2 + 1/1 + \cdots$$

when s takes numerical values not less than 2. Why does this fail for $s = 0$ and $s = 1$?
(c) Determine a solution (x, y) in polynomials with rational coefficients for

$$x^2 - (9s^2 + 12s + 7)y^2 = 1.$$

Exercise 5.9. When $t = 3s + 1$, then $t^2 + 3 = 9s^2 + 6s + 4$. Determine a solution (x, y) in polynomials with rational coefficients for

$$x^2 - (9s^2 + 6s + 4)y^2 = 1.$$

Exercise 5.10. Consider the case $d = t^2 - 4$. For numerical values of t, determine the quantities a_i that occur in the continued fraction expansion of $\sqrt{d}$. Look for patterns and use them to obtain polynomial results.

Exercise 5.11. When $t = 2s + 1$, $t^2 - 4 = 4s^2 + 4s - 3$. Use the continued fraction expansion of $\sqrt{4s^2 + 4s - 3}$ to solve $x^2 - (4s^2 + 4s - 3)y^2 = 1$. For which numerical values of s does this give a valid continued fraction expansion?

Exercise 5.12. When $t = 2s$, $t^2 - 4 = 4s^2 - 4$. Prove that

$$\sqrt{4s^2 - 4} = (2s - 1) + 1/1 + 1/(s - 2) + 1/2(2s - 1) + 1/1 + \cdots.$$

For which numerical values of s does this give a valid continued fraction expansion? Solve

$$x^2 - (4s^2 - 4)y^2 = 1.$$

5.6 Low Values Assumed by $x^2 - dy^2$

The continued fraction algorithm provides us not only with solutions of $x^2 - dy^2 = 1$ but also of $x^2 - dy^2 = k$ for *every* integer k for which $|k| < \sqrt{d}$. Let us see why this is so.

Exercise 6.1.

(a) Suppose a, r, s are positive integers for which $a < \sqrt{d}$ and $r^2 - ds^2 = a$. Prove that

$$0 < r - s\sqrt{d} < \frac{1}{2s}$$

and deduce that

$$\frac{r}{s} - \sqrt{d} < \frac{1}{2s^2}.$$

(b) Suppose b, r, s are positive integers for which $b < \sqrt{d}$ and $r^2 - ds^2 = -b$. Prove that

$$0 < \frac{s}{r} - \frac{1}{\sqrt{d}} < \frac{1}{2r^2}. \quad ♠$$

Let z be a positive irrational number and let r and s be coprime positive integers. Suppose that

$$z = a_0 + 1/a_1 + 1/a_2 + \cdots + a/a_{n-1} + 1/u_n$$

and

$$\frac{r}{s} = b_0 + 1/b_1 + 1/b_2 + \cdots + 1/b_{n-1} + 1/v_n$$

are the developments of z and r/s as finite continued fractions for each positive integer n with a_i and b_i nonnegative integers and u_n and v_n each real numbers at least equal to 1. Note that since r/s is rational, its continued fraction development will terminate, so that $r/s = b_0 + 1/b_1 + \cdots + 1/b_m$. Let its convergents be r_i/s_i $(0 \leq i \leq m)$; observe that $r_1 < r_2 < \cdots < r_m = r$, $s_1 < s_2 < \cdots < s_m = s$.

How close do z and r/s have to be for their convergents up to r/s to agree? In view of our overall goal and Exercise 6.1, we would like this to happen when

$$\left| z - \frac{r}{s} \right| < \frac{1}{2s^2}.$$

Exercise 6.2. Suppose that

$$\sqrt{d} = a_0 + 1/a_1 + 1/a_2 + 1/a_3 + \cdots.$$

What is the continued fraction of $1/\sqrt{d}$? How do the convergents of $1/\sqrt{d}$ compare with the convergents of $\sqrt{d}$?

Exercise 6.3. Let r and s be coprime positive integers for which

$$\left|\sqrt{d} - \frac{r}{s}\right| < \frac{1}{2s^2}.$$

Select the positive integer n for which $q_n \leq s < q_{n+1}$. Suppose that $r/s \neq p_n/q_n$.

(a) From Exercise 3.12, deduce that

$$\left|q_n\sqrt{d} - p_n\right| \leq \left|s\sqrt{d} - r\right|.$$

(b) Prove that

$$\left|\sqrt{d} - \frac{p_n}{q_n}\right| < \frac{1}{2sq_n}.$$

(c) Use the fact that $|sp_n - rq_n| \geq 1$ to prove that

$$\frac{1}{sq_n} \leq \left|\frac{p_n}{q_n} - \frac{r}{s}\right| < \frac{1}{2sq_n} + \frac{1}{2s^2}$$

and then $s < q_n$.

(d) From the contradiction in (c), deduce that $r/s = p_n/q_n$.

(e) What happens if we drop the condition that r and s are coprime?

Exercise 6.4. Prove a result analogous to Exercise 6.3 in the event that

$$\left|\frac{s}{r} - \frac{1}{\sqrt{d}}\right| < \frac{1}{2r^2}.$$

Exercise 6.5. Suppose that $|k| < \sqrt{d}$. Argue that we can obtain every solution in integers x and y to $x^2 - dy^2 = k$ from the continued fraction expansion of $\sqrt{d}$.

5.7 Explorations

Exploration 5.3. A variant on the continued fraction algorithm is to overshoot rather than undershoot at each step. We can formulate it as follows. Start with an irrational number z and let

$$z = b_0 - \frac{1}{v-1},$$

where $b_0 = \lceil x \rceil$ and $v_1 > 1$. Continue as follows

$$v_1 = b_1 - \frac{1}{v_2}, \quad \ldots \quad v_n = b_n - \frac{1}{v_{n+1}},$$

where for each n, $b_n = \lceil v_n \rceil$ and $v_n > 1$. Note that $b_i > 2$. For example, when $z = \sqrt{13}$, the sequence $\{b_n\}$ is $\{4, 3, 3, 2, 2, 2, 2, 2, 3, 3, 8, 3, 3, \ldots\}$. As before, we can define convergents p_n/q_n for the continued fraction. How do these convergents compare to the value of z? Do we still obtain solutions of $x^2 - dy^2 = 1$ for nonsquare values of d from this approach?

Exporation 5.4. Another variant to the continued fraction method is to take the nearest integer at each iteration. Thus, for irrational z,

$$z = c_0 \pm \frac{1}{w_1},$$

where c_0 is the nearest integer and $w_1 > 2$; generally,

$$w_n = c_n \pm \frac{1}{w_{n+1}},$$

where c_n is the nearest integer to w_n and $w_{n+1} > 2$. When $z = \sqrt{13}$, then $\{c_n\}$ is $\{4, 3, 2, 7, 3, 2, \ldots\}$ and the sequence of convergents is $\{4/1, 11/3, 18/5, 137/18, 393/109, 649/180, \ldots\}$. Is this a good method for getting a fundamental solution to Pell's equation?

5.8 Notes and References

The topic of continued fractions and Pell's equation is covered in many number theory textbooks. The following may be consulted:

William J. LeVeque, *Topics in Number Theory, Volume 1.* Addison-Wesley, Reading, MA 1956 (Chapters 8 and 9).

Richard A. Mollin, *Fundamental Number Theory with Applications.* CRC Press, 1998 (Chapters 5 and 6).

Ivan Niven, Herbert S. Zuckerman, and Hugh L. Montgomery, *An Introduction to the Theory of Numbers, 5th edition.* John Wiley, 1991 (Chapter 7).

6.4–6.6. A.M.S. Ramasamy, Polynomial solutions for the Pell's equation. *Indian Journal of Pure and Applied Mathematics* 25 (1994), 577–581.

Exploration 5.1. The method of Tenner is described on page 372 of Volume II of L.E. Dickson, *History of the Theory of Numbers* (Chelsea, NY, 1951).

Exploration 5.3, 5.4. Recent work of A. Mollin explores alternative ways of handling continued fractions to obtain solutions of Pell's equation. See also the survey paper of H.C. Williams.

5.9 Hints

3.6(a). To handle the third equality, multiply numerator and denominator by a_{m+1} and remember the recursion that defines ϕ_n and ϕ_{m-1}.

3.6(b). For the symmetry property, use induction on n.

3.6(c). The proof is similar to the corresponding result for the Euclidean algorithm for a coprime pair of integers.

4.2(d). Note that $A_n = A_n - (Az^2 + Bz + c)q_{n+1}^2$. Try to get an expression involving $p_{n+1} - zq_{n-1}$.

4.3(d). Observe that $(Az^2 + Bz + c)q_{n-1}^2 = 0$ and subtract this quantity from both sides of the equation to determine A_n in (a).

5.6(a). $4t^2 + 4t + 5$ is close to $(2t + 1)^2$.

5.6(b). $4t^4 + 4t + 2$ is close to $(2t^2)^2$.

5.6(c). $16t^4 + 8t^3 + 9t^2 + 6t + 2$ is close to $(4t^2 + t + 1)^2$.

6

Pell's Equation and Pythagorean Triples

As we have seen in Chapter 1, Pell's equation comes up in situations requiring integer solutions to quadratic equations. One well-known equation is that of Pythagoras, $x^2 + y^2 = z^2$; this is closely related to Pell's equation. We continue an investigation begun in Section 1.3.

6.1 A Special Second-Order Recursion

In this section we study the sequence $\{t_n\}$ defined by the following relations

$$t_1 = 1, \qquad t_2 = 6, \qquad t_n = 6t_{n-1} - t_{n-2} \qquad (n \geq 3).$$

The terms of this sequence can be used to generate Pythagorean triples whose smallest numbers differ by 1. Later sections will show how to find other classes of Pythagorean triples.

Exercise 1.1. List the first seven terms of the sequence $\{t_n\}$ in a column. Leave room for other columns of data.

Exercise 1.2. For each n, prove that $t_{n+1} + t_n$ is odd and therefore can be expressed as the sum of two consecutive integers: $t_{n+1} + t_n = a_n + b_n$.

Exercise 1.3. Let $c_n = t_{n-1} - t_n$. Check that $a_n^2 + b_n^2 = c_n^2$ for low values of n. ♠

It is interesting that each t_n can be expressed as a product of corresponding terms of sequences we have already studied. Recall from Chapter 1 the sequences $\{p_n\}$ and $\{q_n\}$ satisfying the recursions

$$p_1 = 1 \quad p_2 = 3 \quad p_n = 2p_{n-1} + p_{n-2} \quad (n \geq 3),$$

$$q_1 = 1 \quad q_2 = 2 \quad q_n = 2q_{n-1} + q_{n-2} \quad (n \geq 3).$$

From Exercise 1.2.4 we have that

$$p_n q_n = 6p_{n-1}q_{n-1} - p_{n-2}q_{n-2} \ (n \geq 3).$$

Exercise 1.4. Prove by induction that $t_n = p_n q_n$ for $n \geq 3$.

Exercise 1.5. Recall that the recursions for $\{p_n\}$ and $\{q_n\}$ satisfy

$$p_n = p_{n-1} + 2q_{n-1} \quad \text{and} \quad q_n = p_{n-1} + q_{n-1}$$

for $n \geq 2$. Prove that:
(a) $p_n = 6p_{n-2} - p_{n-4}$ and $q_n = 6q_{n-2} - q_{n-4}$ for $n \geq 5$.
(b) $p_{2n+1} = a_n + b_n$ and $q_{2n+1} = c_n$ for $n \geq 1$.
(c) $q_{2n} = 2p_n q_n$ for $n \geq 1$.
(d) $p_n^2 + p_{n+1}^2 = 2q_{2n+1}$ and $q_n^2 + q_{n+1}^2 = q_{2n+1}$ for $n \geq 1$.

Exercise 1.6. Consider the sequence $\{s_n\}$ defined by

$$s_1 = 3, \quad s_2 = 17, \quad s_n = 6s_{n-1} - s_{n-2}, \quad (n \geq 3).$$

Verify that $s_n = p_{2n}$ and that the Pell's equation $s_n^2 - 8t_n^2 = 1$ holds for small values of n. Try to establish it for general n.

Exploration 6.1. With the notation of Exercise 1.3, does $a_n^2 + b_n^2 = c_n^2$ for each value of n? Do the (a_n, b_n, c_n) give a complete set of solutions of $x^2 + y^2 = z^2$ with $y - x = 1$?

5.2 A General Second-Order Recursion

A second-order recursion is a sequence $\{t_n\}$ whose terms satisfy, for integers n,

$$t_n = \alpha t_{n-1} + \beta t_{n-2} \tag{1}$$

for constant multipliers α, β independent of n. Examples are the sequences $\{t_n\}$, $\{p_n\}$, $\{q_n\}$, and $\{s_n\}$ encountered in Section 1. In this section we will suppose that α and β are integers, and that n ranges over all the integers.

Exercise 2.1. When $\beta = 0$ in (1), the recursion becomes $t_n = \alpha t_{n-1}$. For how many values of n must t_n be known to determine a solution? Suppose t_0 is known. What is t_n?

Exercise 2.2. When $\alpha = 0$, the recursion becomes $t_n = \beta t_{n-2}$. Suppose that t_0 is known. Which values of t_n are determined? What are they? How much additional information is needed to determine the sequence $\{t_n\}$ completely? Experiment with the case $\beta = 1$, $\beta = 4$, $\beta = -1$.

Exercise 2.3. Let α and β both be nonzero. Show that the recursion (1) is satisfied by $t_n = r^n$ for some constant r if and only if r is a root of the quadratic equation

$$x^2 = \alpha x + \beta, \quad \text{or} \quad x^2 - \alpha x - \beta = 0. \quad ♠$$

The expression $x^2 - \alpha x - \beta$ is called the *characteristic polynomial* of the recursion.

Exercise 2.4.

(a) Suppose that $\{t'_n\}$ and $\{t''_n\}$ both satisfy the recursion (1). Verify that

$$t_n = \gamma t'_n + \delta t''_n$$

is also a solution of (1) for any constants γ and δ.

(b) Suppose that the characteristic polynomial $x^2 - \alpha x - \beta$ has two distinct roots r and s. Deduce that $t_n = \gamma r^n + \delta s^n$ satisfies (1) for any constants γ and δ. (We will show in the next exercise that this picks up all solutions to the recursion.)

(c) Suppose that the polynomial $x^2 - \alpha x - \beta$ has a double root r, so that $4\beta = -\alpha^2$ and $r = \alpha/2$. Verify for each n that

$$t_n - rt_{n-1} = r(t_{n-1} - rt_{n-2}),$$

so that $\{t_n/r^n\}$ is an arithmetic progression. Deduce that

$$t_n = (\gamma n + \delta)r^n$$

for some constants γ and δ.

Exercise 2.5. Suppose that $\{t_n\}$ is any solution of (1), and that the characteristic polynomial has two distinct roots r and s. Verify that the system

$$\gamma + \delta = t_0,$$
$$\gamma r + \delta s = t_1,$$

is uniquely solvable for the parameters γ and δ. Prove that for these values of γ and δ, $t_n = \gamma r^n + \delta s^n$ for all n.

Exercise 2.6. Establish a result analogous to Exercise 2.5 in the event that the characteristic polynomial has two equal roots.

Exercise 2.7. Use the theory of the previous exercise to write the general terms of the following recursions of Section 1 in the form $t_n = \gamma r^n + \delta s^n$. Check that the formula you obtain works when $n = 1, 2, 3, 4$:

(1) $\{1, 6, 35, 204, \ldots\}$,
(2) $\{1, 2, 5, 12, 29, \ldots\}$,
(3) $\{1, 3, 7, 17, 41, \ldots\}$,
(4) $\{3, 17, 99, 577, \ldots\}$.

Another approach to solving the recursion (1) is through matrices. In this formulation, the recursion is revealed as an analogue of a geometric sequence, in which each term depends on only one predecessor. We can write

$$\begin{pmatrix} t_n \\ t_{n+1} \end{pmatrix} = \begin{pmatrix} 0 & 1 \\ \beta & \alpha \end{pmatrix} \begin{pmatrix} t_{n-1} \\ t_n \end{pmatrix},$$

so that

$$\begin{pmatrix} t_n \\ t_{n+1} \end{pmatrix} = \begin{pmatrix} 0 & 1 \\ \beta & \alpha \end{pmatrix}^n \begin{pmatrix} t_0 \\ t_1 \end{pmatrix}$$

for each integer n.

Exercise 2.8.

(a) Verify that if r and s are the roots of the quadratic polynomial $x^2 - \alpha x - \beta$, then

$$\begin{pmatrix} 0 & 1 \\ \beta & \alpha \end{pmatrix} \begin{pmatrix} 1 \\ r \end{pmatrix} = r \begin{pmatrix} 1 \\ r \end{pmatrix}$$

and

$$\begin{pmatrix} 0 & 1 \\ \beta & \alpha \end{pmatrix} \begin{pmatrix} 1 \\ s \end{pmatrix} = s \begin{pmatrix} 1 \\ s \end{pmatrix}.$$

Thus, on the vectors $\binom{1}{r}$ and $\binom{1}{s}$, operating with the matrix behaves like a simple multiplication by a scalar.

(b) Suppose that t_0 and t_1 are given and that r and s are distinct. Prove that there are numbers γ and δ such that

$$t_0 = \gamma + \delta \quad \text{and} \quad t_1 = \gamma r + \delta s$$

and thus

$$\begin{aligned} \begin{pmatrix} 0 & 1 \\ \beta & \alpha \end{pmatrix}^n \begin{pmatrix} t_0 \\ t_1 \end{pmatrix} &= \gamma \begin{pmatrix} 0 & 1 \\ \beta & \alpha \end{pmatrix}^n \begin{pmatrix} 1 \\ r \end{pmatrix} + \delta \begin{pmatrix} 0 & 1 \\ \beta & \alpha \end{pmatrix}^n \begin{pmatrix} 1 \\ s \end{pmatrix} \\ &= \gamma \begin{pmatrix} r^n \\ r^{n+1} \end{pmatrix} + \delta \begin{pmatrix} s^n \\ s^{n+1} \end{pmatrix}. \end{aligned}$$

Note that this corroborates the result of Exercise 2.6.

Exercise 2.9. Let

$$A = \begin{pmatrix} 0 & 1 \\ \beta & \alpha \end{pmatrix}, \quad M = \begin{pmatrix} 1 & 1 \\ r & s \end{pmatrix}, \quad \text{and} \quad N = \frac{1}{s-r} \begin{pmatrix} s & -1 \\ -r & 1 \end{pmatrix},$$

where r and s are distinct roots of $x^2 - \alpha x - \beta$.

(a) Verify that $A^2 - \alpha A - \beta I = O$, where I is the identity matrix $\begin{pmatrix} 1 & 0 \\ 0 & 1 \end{pmatrix}$. (Cf. Exercise 1.2.3.)

(b) Verify that $NM = MN = I$, where I is the identity matrix $\begin{pmatrix} 1 & 0 \\ 0 & 1 \end{pmatrix}$.

(c) Prove that $NA^nM = (NAM)^n$ for each positive integer n.

(d) Verify that

$$NAM = \begin{pmatrix} r & 0 \\ 0 & s \end{pmatrix}$$

and deduce that

$$NA^nM = \begin{pmatrix} r^n & 0 \\ 0 & s^n \end{pmatrix}. \quad ♠$$

The significance of the result of Exercise 2.9(c) is that $A \to NAM$ is a "similarity" transformation of the matrix A into a form where it becomes apparent that A is a type of amalgam of multiplication by r and multiplication by s, so that the second-order recursion becomes a sort of two-story geometric sequence.

Exercise 2.10. Let $\{t_n\}$ be a general second order recursion defined by (1). Prove that

$$t_n^2 - t_{n+1}t_{n-1} = -\beta(t_{n-1}^2 - t_n t_{n-2}).$$

Exercise 2.11. Let $r_n = \alpha t_n + 2\beta t_{n-1} = t_{n+1} + \beta t_{n-1}$ for $n \geq 1$, and let $D = \alpha^2 + 4\beta$ be the discriminant of the characteristic polynomial $x^2 - \alpha x - \beta$. Prove that

$$r_n^2 - Dt_n^2 = -4\beta(t_n^2 - t_{n+1}t_{n-1}) = (-1)^n(t_1^2 - t_2 t_0)4\beta^n \quad (n \geq 1).$$

Exercise 2.12. Consider the special case $\beta = -1$, $t_0 = 0$, $t_1 = 1$, so that

$$t_n = \alpha t_{n-1} - t_{n-2} \quad (n \in \mathbf{Z}),$$
$$D = \alpha^2 - 4,$$
$$r_n = \alpha t_n - 2t_{n-1} = t_{n+1} - t_{n-1}.$$

(a) Write out the values of r_n and t_n as functions of α for $n = -1, 0, 1, 2, 3, 4$.
(b) Prove that $r_n = \alpha r_{n-1} - r_{n-2}$ and that $r_n^2 - Dt_n^2 = 4$.

Exercise 2.13. In Exercise 2.12, let $\alpha = 2\gamma + 1$ be odd. Verify that $r_n \equiv t_n \equiv 0 \pmod 2$ if and only if n is a multiple of 3. This permits us to find certain solutions to the Pell's equation $x^2 - (\alpha^2 - 4)y^2 = 1$. For example, taking $n = 3$, check that we obtain

$$(4\gamma^3 + 6\gamma^2 - 1)^2 - (4\gamma^2 + 4\gamma - 3)(2\gamma^2 + 2\gamma)^2 = 1.$$

Exercise 2.14. In Exercise 2.11, let $\alpha = 2\gamma$ be even. Define $s_n = \frac{1}{2}r_n = \gamma t_n - t_{n-1}$.

Verify that $(x, y) = (s_n, t_n)$ satisfies the equation

$$x^2 - (\gamma^2 - 1)y^2 = 1.$$

Cf. Exercise 3.4.1. We have that

$$s_n + \left(\sqrt{\gamma^2 - 1}\right)t_n = \left(\gamma + \sqrt{\gamma^2 - 1}\right)^n.$$

Exercise 2.15. In Exercise 2.10, let $\beta = 1$, $t_0 = 0$, $t_1 = 1$, so that

$$t_n = \alpha t_{n-1} + t_{n-2} \quad (n \in \mathbf{Z})$$
$$D = \alpha^2 + 4,$$
$$r_n = \alpha t_n + 2t_{n-1} = t_{n+1} + t_{n-1} \quad (n \in \mathbf{Z}).$$

(a) Verify that $r_n^2 - Dt_n^2 = (-1)^n 4$.

(b) Show that, if $\alpha = 2\gamma$, $s_n = \frac{1}{2}r_n$, then $(x, y) = (s_n, t_n)$ satisfies
$$x^2 - (\gamma^2 + 1)y^2 = \pm 1.$$

(c) What are $\{s_n\}$ and $\{t_n\}$ when $\gamma = 1$?

Exercise 2.16. Consider the case $\alpha = 3$, $\beta = -1$, $t_0 = 1$, $t_1 = 5$, so that
$$t_n = 3t_{n-1} - t_{n-2},$$
$$D = 5,$$
$$r_n = 3t_n - 2t_{n-1} = t_{n+1} - t_{n-1}.$$
Determine the terms r_n and t_n for positive and negative values of n using the recursion equation to extrapolate both forwards and backwards from $n = 0$ and $n = 1$. Obtain integer solutions to the Pell's equation
$$x^2 - 5y^2 = 44.$$

Exploration 6.2. Does this method ever provide a complete set of solutions to $x^2 - dy^2 = k$?

5.3 Pythagorean Triples

The example in Section 6.1 gives an illustration of how to find solutions to a system of Diophantine equations of the form
$$rx + sy = w,$$
$$rx^2 + sy^2 = z^2,$$
$$y - x = e$$
where r, s, e are given coefficients. In that example we had $r = s = e = 1$ and $\{x, y\} = \{a_n, b_n\}$. In the solutions, w was the sum of consecutive terms in a recursion, and z was their difference.

To set this up more generally, let us find conditions on the multipliers α and β of a second-order recursion and on the coefficients r and s to make such a solvable system possible. Let a, b, and $\alpha b + \beta a$ be three consecutive terms of a second-order recursion. We want to ensure the existence of u and v for which
$$ru + s(u + e) = a + b, \tag{1}$$
$$rv + s(v + e) = b + (\alpha b + \beta a), \tag{2}$$
$$ru^2 + s(u + e)^2 = (b - a)^2, \tag{3}$$
$$rv^2 + s(v + e)^2 = (\alpha b + \beta a - b)^2. \tag{4}$$

Exercises 3.1. Suppose (1) holds. Subtract (1) from (2) to obtain
$$v = u + \frac{\alpha b + (\beta - 1)a}{r + s}. \tag{5}$$

Exercise 3.2. We eliminate u and v from the system (1)–(4). Substitute (5) into (4) and take account of (1) and (3) to get

$$[\alpha b + (\beta - 1)a]^2 + 2[\alpha b + (\beta - 1)a][a + b]$$
$$= k\big[(\alpha^2 - 2\alpha)b^2 + 2(\alpha\beta - \beta + 1)ab + (\beta^2 - 1)a^2\big] \tag{6}$$

where $k = r + s$.

Exercise 3.3. We want to make (6) hold for any three consecutive terms of a recursion with multipliers α and β. Accordingly, (6) should be an identity in a and b. Equate coefficients of b^2, ab and a^2 in both sides to obtain

$$\alpha^2 + 2\alpha = k(\alpha^2 - 2\alpha), \tag{7}$$
$$\alpha\beta + \beta - 1 = k(\alpha\beta - \beta + 1), \tag{8}$$
$$\beta^2 - 1 = k(\beta^2 - 1). \tag{9}$$

Exercise 3.4. One obvious possibility for the system (7)–(9) is $\alpha = 0$, $\beta = 1$. This corresponds to the sequence $a, b, a, b, a, b, \ldots$. In this case, $u = v$, and we have to satisfy $rx + sy = a + b$, $rx^2 + sy^2 = (b - a)^2$ for integers x and y where r and s are given, a system that may or may not be solvable in integers.

(a) Suppose that $\alpha = 0$ and $\beta \neq 1$. Show that (7)–(9) require that $k = \beta = -1$ and that the three consecutive terms be $a, b, -a$. Verify that $v = 2a + u$ and that any choice of r, s, u, v satisfying (1) and (3) will also give a solution to (2) and (4).

(b) Suppose $\alpha = -2$. Show that $k = 0$ and $\beta = -1$ are needed to satisfy (7)–(9), and that this corresponds to the case $r = -s$ and the sequence $a, b, a - 2b$. Verify in this case that (1) and (2) are not consistent unless $a + b = 0$.

Exercise 3.5. Suppose that α and k are both nonzero.

(a) Deduce from (7) that $\alpha \neq 2$ and $k \neq 1$, so that $\beta^2 = 1$.

(b) Use (7) and (8) to deduce that

$$(\alpha + 2)(\alpha\beta - \beta + 1) = (\alpha - 2)(\alpha\beta + \beta - 1)$$

whence $\alpha(\beta + 1) = 0$. Deduce that $\beta = -1$ and $k = \frac{\alpha+2}{\alpha-2}$.

Exercise 3.6. Deduce from the foregoing exercises the following result:

Suppose that $\{t_n : n = 0, \pm 1, \pm 2, \ldots\}$ is a sequence for which

$$t_1 = a, \quad t_2 = b, \quad t_n = \alpha t_{n-1} - t_{n-2}, \quad (n \in \mathbf{Z})$$

where $\alpha \neq 2$. Let $k = \frac{\alpha+2}{\alpha-2}$, r, and s be such that $k = r + s$ and the system

$$rx + sy = w,$$
$$rx^2 + sy^2 = z^2,$$
$$y - x = e,$$

has a solution (x_0, y_0, z_0, w_0) with $z_0 = t_1 - t_0$ and $w_0 = t_1 + t_0$, where e is a constant. Then there are additional solutions (x_n, y_n, z_n, w_n) to the system with

$$\begin{aligned} z_n &= t_{n+1} - t_n, \\ w_n &= t_{n+1} + t_n, \\ x_n &= x_{n-1} + \frac{\alpha t_n - 2t_{n-1}}{k}, \\ y_n &= x_n + e. \quad \spadesuit \end{aligned}$$

The example in Section 1 corresponds to the case $(e, r, s, k, \alpha) = (1, 1, 1, 2, 6)$.

Exercise 3.7. We can determine Pythagorean triples (x, y, z) for which $y - x = 7$. In this case, show that we ought to take $e = 7$, $r = s = 1$, $k = 2$, and $\alpha = 6$. Noting that one such triple is $(-4, 3, 5)$, we select a and b such that $a + b = (-4) + 3 = -1$ and $b - a = 5$. Embed these values of a and b as adjacent terms in a suitable recursion and generate infinitely many solutions of the system

$$\begin{aligned} x + y &= w, \\ x^2 + y^2 &= z^2, \\ y - x &= 7. \end{aligned}$$

Exercise 3.8. Let κ be an arbitrary parameter. The first two terms of the Pythagorean triple $(2\kappa + 1, 2\kappa^2 + 2\kappa, 2\kappa^2 + 2\kappa + 1)$ differ by $2\kappa^2 - 1$. Generate infinitely many solutions of the system

$$\begin{aligned} x + y &= w, \\ x^2 + y^2 &= z^2 \\ y - x &= 2\kappa^2 - 1. \end{aligned}$$

Exercise 3.9. Let m and n be parameters. The first two terms of the Pythagorean triple $(m^2 - n^2, 2mn, m^2 + n^2)$ differ by $2mn + n^2 - m^2$. Verify that the sequence t_n used to generate other triples whose smallest two numbers have the same difference contains the consecutive terms

$$\begin{aligned} mn - n^2 &= (m - n)n, \\ m^2 + mn &= (m + n)m, \\ 6m^2 + 5mn + n^2 &= (3m + n)(2m + n), \\ 35m^2 + 29mn + 6n^2 &= (7m + 3n)(5m + 2n). \end{aligned}$$

Exercise 3.10. Consider the Diophantine equation $x^2 + 2y^2 = z^2$.

(a) What values of r, s, k, and α should be used in generating solutions?

(b) Consider the sequence $\{1, 4, 15, 56, 209, \ldots\}$. We create the following table:

z_n	t_n	w_n	equation
	1		
3		$5 = 1 + 2 \cdot 2$	$3^2 = 1^2 + 2 \cdot 2^2$
	4		
11		$19 = 7 + 2 \cdot 6$	$11^2 = 7^2 + 2 \cdot 6^2$
	15		
41		$71 = 23 + 2 \cdot 24$	$41^2 = 23^2 + 2 \cdot 24^2$
	56		
153		$265 = 89 + 2 \cdot 88$	$153^2 = 89^2 + 2 \cdot 88^2$
	209		

If we look carefully, we observe that the $y - x$ difference takes the alternate values $+1$ and -1. Readjust to get two tables, one for $y - x = 1$ always occurring and the other for $y - x = -1$ always occurring. In these new tables you will find that x and y will not always be integers.

Exercise 3.11. To explain the ambiguity arising in Exercise 3.10, suppose that we already know a, b, r, s, and k. Then u and e have to satisfy $ku + se = a + b$ and $ru^2 + s(u + e)^2 = (b - a)^2$. Eliminate u from these equations to obtain

$$rse^2 = (k - 1)(a^2 + b^2) - 2(k + 1)ab.$$

Argue that there may be two different possible values of e, one the negative of the other, that can be used for the given values of the remaining parameters. However, note that the ambiguity is masked when $r = s$; why is this?

Exercise 3.12. Show that, corresponding to the diophantine equation $x^2 + 3y^2 = z^2$, the solution $(x, y, z) = (1, 4, 7)$ leads to the sequence $\{\ldots, 3, 10, 91/3, \ldots\}$. Generate other solutions in integers to the equation $x^2 + 3y^2 = z^2$.

Exercise 3.13. Starting with the solution $(x, y, z) = (4, 5, 7)$ to the equation $3x^2 + 2y^2 = 2z^2$, generate other solutions in integers to this equation.

6.4 A Method of Euler

Over two centuries ago, Leonhard Euler was able to exploit Pythagorean triples and some analogues of them to generate solutions to various Pell's equations. Here is how he did it.

Exercises 4.1.

(a) Suppose that $(x, y) = (q, p)$ is a solution of $x^2 - dy^2 = -1$. Verify that $(x, y) = (2q^2 + 1, 2pq)$ satisfies $x^2 - dy^2 = 1$.

(b) Thus, from values of p and q for which $dp^2 - 1 = q^2$, we can generate solutions to $x^2 - dy^2 = 1$. Our standpoint is to fix a value of p and find solutions of Pell's equation for integers of the form $d = (q^2+1)/p^2$. We observe that q^2+1 is (trivially) a sum of two squares and that the product of the sum of two squares is also a sum of two squares. So we ask the same of p^2; thus, p is the largest integer of a Pythagorean triple.

Suppose that $p^2 = b^2 + c^2$. In order for d to be an integer, we ask that

$$q^2 + 1 = (b^2 + c^2)(f^2 + g^2) = (bf + cg)^2 + (bg - cf)^2.$$

Given p, b, c, we determine solutions (g, f) of $bg - cf = \pm 1$ and then set $q = bf + cg$. Beginning with $5^2 = 3^2 + 4^2$, explain how Euler arrived at the following table and check in each case that $x^2 - dy^2 = 1$.

f	1	2	4	5	7	8
g	1	3	5	7	9	11
q	7	18	32	43	57	68
d	2	13	41	74	130	185
x	99	649	2049	3699	6499	9249
y	70	180	320	430	570	680

(c) Take $p = 13$ and use Euler's method to determine solutions to Pell's equations.
(d) Take $p = 17$ and use Euler's method to determine solutions to Pell's equations.

Exercise 4.2.
(a) Suppose that $(x, y) = (q, p)$ satisfies $x^2 - dy^2 = -2$. Verify that $(x, y) = (q^2 + 1, pq)$ satisfies $x^2 - dy^2 = 1$.
(b) In this case, we find that $d = (q^2 + 2)/p^2$, so that we try to arrange that both p^2 and $q^2 + 2$ have the form $x^2 + 2y^2$. Verify that the set of numbers of this form is closed under multiplication.
(c) Start with $p^2 = b^2 + 2c^2$ and

$$q^2 + 2 = (b^2 + 2c^2)(f^2 + 2g^2) = (bf + 2cg)^2 + 2(cf - bg)^2,$$

taking $cf - bg = \pm 1$ and $bf + 2cg = q$. Explain how Euler might have completed a table analogous to that of Exercise 4.1 of the case $p = 3$.

Exercise 4.3. Other starting relations for p and q used by Euler were $dp^2 + 2 = q^2$, $dp^2 + 4 = q^2$, and $dp^2 - 4 = q^2$. Explain, in each case, how the pair (p, q) can be used to generate solutions to certain Pell's equations $x^2 - dy^2 = 1$ and imitate the method of Exercises 4.1 and 4.2.

6.5 Notes

Section 4: The method described is in the paper Leonhard Euler, *Nova subsida pro resolutione formulae axx + 1 = yy*. *Opuscula analytica* 1 (1783), 310–328 = *Opera Omnia* (1) 4, 76–90 (Enestrom No. 559).

6.6 Hints

2.5. Use the fact that $\{t_n\}$ and $\{\gamma r^n + \delta s^n\}$ satisfy the same recursion along with an induction argument.

2.8(b). If M is a matrix, $\mathbf{v}$ and $\mathbf{w}$ are vectors, and λ and μ are scalars, then $M(\lambda\mathbf{v} + \mu\mathbf{w}) = \lambda M(\mathbf{v}) + \mu M(\mathbf{w})$.

2.9. Start with $n = 2$, and note that $NA^2M = NAMNAM$.

3.12. Note that $x^2 + 3y^2 = z^2$ is homogeneous, so that any integer multiple of a solution is also a solution. Thus, we can generate integer solutions from rational solutions.

3.13. Consider the equation $\frac{3}{2}x^2 + y^2 = z^2$.

7

The Cubic Analogue of Pell's Equation

In looking for a higher-degree version of Pell's equation, a natural choice is the equation

$$x^3 - dy^3 = k,$$

where d is a noncubic integer. However, it turns out that the solutions of such equations are not very numerous, nor do they exhibit the nice structure found in the quadratic case. This chapter will begin with an investigation of this type before treating a better analogue that admits a theory comparable to the quadratic version. This is the equation:

$$x^3 + cy^3 + c^2z^3 - 3cxyz = k,$$

where c is any integer other than a perfect cube and k is an integer. In this chapter we will see how this equation comes about and examine its theory. It will turn out, as in the quadratic case, that there is a fundamental solution; however, this solution is not so neatly obtained and some ad hoc methods are needed.

Sections 6.2 and 6.3 can be skipped, since they are independent of the rest of the chapter.

6.1 The Equation $x^3 - dy^3 = 1$: Initial Skirmishes

Exercise 1.1. Suppose $d = r^3$, a cubic integer. By using the factorization $x^3 - r^3y^3 = (x - ry)(x^2 + rxy + r^2y^2)$, prove that any solution (x, y) in integers of $x^3 - dy^3 = \pm 1$ must satisfy $x^2 + rxy + r^2y^2 = 1$. Use this fact to deduce that either x or y must vanish and so obtain all possible solutions of $x^3 - r^3y^3 = \pm 1$.

Exercise 1.2.

(a) Verify that $x^3 - dy^3 = 1$ has a solution with x and y both nonzero when d has the form $s^3 - 1$ for some integer s.

(b) Try to determine solutions of $x^3 - 7y^3 = 1$ other than $(x, y) = (1, 0), (2, 1)$.

Exercise 1.3.

(a) Prove that $x^3 - dy^3 = 1$ has a solution for which $y = 2$ if and only if $d = u(64u^2 + 24u + 3)$ for some integer u.

(b) Determine a solution in nonzero integers of $x^3 - 43y^3 = 1$.

Exercise 1.4.

(a) Prove that $x^3 - dy^3 = 1$ has a solution for which $y = 3$ if and only if $d = u(27u^2 + 9u + 1)$.

(b) Determine a solution in nonzero integers for $x^3 - 19y^3 = 1$ and for $x^3 - 37y^3 = 1$.

Exercise 1.5. Suppose that x and y are nonzero intgers for which $x^3 - dy^3 = 1$. This equation can be rewritten as $dy^3 = (x-1)(x^2+x+1)$.

(a) Verify that the greatest common divisor of $x - 1$ and $x^2 + x + 1$ is equal to 3 when $x \equiv 1 \pmod 3$ and to 1 otherwise.

(b) Prove that $x^2 + x + 1$ is never divisible by 9.

(c) Deduce from (a) and (b) that $x^2 + x + 1$ must be of the form rv^3 or $3rv^3$, where r is an odd divisor of d.

Exercise 1.6. Suppose in Exercise 1.5 that $x^2 + x + 1 = v^3$.

(a) Give a numerical example to show that this equation actually has a solution for which x and v are both nonzero.

(b) Find nonzero integers x and y for which $x^3 - 17y^3 = 1$.

Exercise 1.7. Suppose in Exercise 1.5 that $x^2 + x + 1 = 3v^3$, so that $x \equiv 1 \pmod 3$. Setting $x = 3u + 1$, obtain the equation $(u+1)^3 = u^3 + v^3$ and obtain from Fermat's last theorem that this case is not possible. (See **Exploration 7.1.**)

Exercise 1.8. Make a table of some integers d with $2 \le d \le 100$ for which $x^3 - dy^3 = 1$ has at least one solution with $xy \neq 0$. List the solutions. Did you find any values of d for which there are two such solutions?

7.2 The Algebraic Integers in $Q(\sqrt{-3})$

We have seen that $x^3 - dy^3 = 1$ is nontrivially solvable for certain values of d. Rewriting this equation as $x^3 + (-1)^3 = dy^3$, we see that we have to study equations of the form $x^3 + z^3 = dy^3$. One way to approach this is to factor the left side as $(x+z)(x+\omega z)(x+\omega^2 z)$, where $\omega = \frac{1}{2}(-1+\sqrt{-3})$ is an imaginary cube root of unity; thus, $\omega^3 = 1$ and $\omega^2 + \omega + 1 = 0$. This can be treated as a factorization in the quadratic field $\mathbf{Q}(\sqrt{-3}) \equiv \{p+q\sqrt{-3} : p, q \in \mathbf{Q}\}$ treated in Section 4.3. The set $\mathbf{Q}(\sqrt{-3})$ is closed under addition, subtraction, multiplication, and division by nonzero elements. The norm $N(a+b\sqrt{-3})$ of numbers of the form $a + b\sqrt{-3}$ is defined by the product of the number and its surd conjugate $a - b\sqrt{-3}$, namely $a^2 + 3b^2$. Note that the norm is always nonnegative.

Recall that a number θ is an algebraic integer in $\mathbf{Q}(\sqrt{-3})$ if and only if it is a root of a quadratic equation of the form $t^2 + bt + c = 0$, where b and c are integers. Normally, the reciprocal of an algebraic integer is not an algebraic integer. An algebraic integer θ is a *unit* if θ and its reciprocal $1/\theta$ are both algebric integers. The set of algebraic integers in $\mathbf{Q}(\sqrt{-3})$ will be denoted by I. This set contains the ordinary integers.

Exercise 2.1.

(a) Explain why $\omega^2 + \omega + 1 = 0$.

(b) Verify that $\sqrt{-3} = \omega - \omega^2 = 1 + 2\omega$.

(c) Prove that $\mathbf{Q}(\sqrt{-3})$ is the set of numbers of the form $r + s\omega$, where r and s are rational.

(d) Prove that the surd conjugate of ω is ω^2, and thus the surd conjugate of $r + s\omega$ is $r + s\omega^2$.

Exercise 2.2. Let α and β be members of $\mathbf{Q}(\sqrt{-3})$. Prove that $N(\alpha\beta) = N(\alpha)N(\beta)$.

Exercise 2.3. Let $\theta = r + s\omega$. Verify that its reciprocal $1/\theta$ is equal to $(r + s\omega^2)/(r^2 - rs + s^2)$.

Exercise 2.4.

(a) Suppose that θ belongs to I. Prove that $N(\theta)$ is an ordinary nonnegative integer and vanishes only when $\theta = 0$.

(b) Prove that θ is a unit if and only if $N(\theta) = 1$.

(c) Suppose that $r + s\omega$ is a unit, so that $r^2 - rs + s^2 = 1$. Prove that $2r - s$ is an ordinary integer. Deduce that r and s are integers for which $(2r - s)^2 + 3s^2 = 4$, so that $(r, s) = (\pm 1, 0), (0, \pm 1), (\pm 1, \pm 1)$.

(d) Prove that the only units in I are ± 1, $\pm\omega$ and $\pm\omega^2$. ♠

Analyzing the equation $x^3 + z^3 = dy^3$ involves examining the factorization of $x^3 + z^3$. In the following exercises we will develop a theory of factorization for the system I that is similar to that for ordinary integers. This will involve notions of divisibility and primality as well as a version of the fundamental theorem of arithmetic that allows each algebraic integer to be decomposed as a product of primes. The development will involve an adaptation of the Euclidean algorithm.

Let α and β be members of I. We say that $\beta|\alpha$ ("β divides α") if $\alpha = \beta\gamma$ for some γ in I. Every member of I divides 0 and is divisible by each of the units. We say that ρ in I is prime if $N(\rho) \neq 1$ and it is divisible only by numbers of the form ϵ and $\epsilon\rho$, where ϵ is a unit. We will see that an ordinary prime integer may or may not be a prime in I.

Exercise 2.5.
(a) Suppose that ρ belongs to I and that $N(\rho)$ is an ordinary prime integer. Prove that ρ must be prime.
(b) Deduce from (a) that $1 - \omega$ is a prime.

Exercise 2.6. We show that although $N(2)$ is composite, 2 is actually prime in I. This demonstrates that the converse of Exercise 2.5(a) is not true.
(a) Prove that $N(2) = 4$.
(b) Suppose that $2 = (p+q\omega)(r+s\omega)$ is a product of two members of I, neither of which is a unit. Prove that $p^2 - pq + q^2 = \pm 2$, so that $(2p-q)^2 + 3q^2 = \pm 8$. Show that this equation has no solutions in integers p and q and deduce from the contradiction that 2 must be prime in I.

Exercise 2.7. Verify that $3 = -[\omega(1-\omega)]^2$, so that 3 is not prime in I (although it is an ordinary prime).

Exercise 2.8. Let α and β be two members of I. We can write α/β in the form $u + v\omega$, where u and v are rational (check that this can be done!). Suppose that m and n are integers for which $|u - m| \leq \frac{1}{2}$ and $|v - n| \leq \frac{1}{2}$.
(a) Verify that $N((u-m) + (v-n)\omega) \leq \frac{3}{4}$.
(b) Let $\gamma = m + n\omega$ and $\delta = \alpha - \beta\gamma$. Verify that $\alpha = \gamma\beta + \delta$ and that $N(\delta) \leq \frac{3}{4}N(\beta) < N(\beta)$. ♠

The result of Exercise 2.7(b) is the analogue of the ordinary division algorithm in which we divide a number β into α and get a quotient γ and a remainder δ "smaller" than the divisor. In this case, the measure of size of a number is not the absolute value but the norm. We can now set up the Euclidean algorithm in the same way as for ordinary integers. Suppose that we are given two algebraic integers α and β. By repeating division, we can obtain

$$\alpha = \gamma\beta + \beta_1,$$
$$\beta = \gamma_1\beta_1 + \beta_2,$$
$$\beta_1 = \gamma_2\beta_2 + \beta_3,$$

and so on, where $N(\beta) > N(\beta_1) > N(\beta_2) > N(\beta_3) > \cdots \geq 0$. Since the norms constitute a decreasing sequence of nonnegative integers, the process cannot go on forever, and we will eventually arrive at an exact division:

$$\beta_{k-2} = \gamma_{k-1}\beta_{k-1} + \beta_k,$$
$$\beta_{k-1} = \gamma_k\beta_k,$$

where β_k is nonzero.

Exercise 2.9.
(a) Prove that β_k must be a common divisor of α and β.
(b) Suppose δ is a divisor of α and β. Prove that δ is a divisor of β_k. ♠

Let α and β belong to I. A *greatest common divisor* of α and β is an element of I that divides both α and β, and in turn is divisible by every common divisor of α and β. The number β_k produced by the Euclidean algorithm in Exercise 2.8 is a greatest common divisor of α and β. We say that α and β are *coprime* if the only common divisors of α and β are units.

Exercise 2.10. Let α, β belong to I. Suppose ρ and σ are two greatest common divisors of α and β. Prove that $\rho = \sigma\epsilon$ for some unit ϵ.

Exercise 2.11.

(a) Suppose that the Euclidean algorithm of Exercise 2.8 is carried out. Observe that

$$\beta_{i+1} = \beta_{i-1} - \gamma_i \beta_i$$

for $1 \leq i \leq k-1$ (where $\beta_0 = \beta$). Use these facts to show that β_k can be written in the form $\xi\alpha + \eta\beta$, where ξ and η belong to I.

(b) Prove that every greatest common divisor of α and β can be written in this form.

Exercise 2.12. Let α and β be a coprime pair in I.

(a) Prove that there are numbers $\xi, \eta \in I$ for which

$$1 = \xi\alpha + \eta\beta.$$

(b) Let $\mu \in I$ and $\beta | \alpha\mu$. Prove that $\beta | \mu$. ♠

With these results in hand, we are in a position to prove the fundamental theorem of arithmetic for I, that, up to units, each member of I can be uniquely written as a product of primes in I.

Exercise 2.13.

(a) Let α be a nonprime member of I. Prove that α can be written as a product $\beta\gamma$ where the norms $N(\beta)$ and $N(\gamma)$ are strictly less than $N(\alpha)$. Extend this result to show that α can be written as a product $\rho_1\rho_2\cdots\rho_k$ of primes.

(b) Suppose that $\alpha = \rho_1\rho_2\cdots\rho_k = \sigma_1\sigma_2\cdots\sigma_l$ are two representations of α as a product of primes in I. Use Exercise 2.12(b) to show that each ρ_i must divide one of the σ_j, so that $\rho_i = \epsilon\sigma_j$ for some unit ϵ.

(c) Prove, in (b), that $k = l$ and that the primes ρ_i and σ_j can be paired so that each ρ_i is the product of σ_j and a unit.

Exercise 2.14. Suppose that $\alpha = \delta^k$ for some positive integer k and for some δ in I and that $\alpha = \beta\gamma$, where all the common divisors of β and γ are units. Prove that β and γ must both be kth powers, up to a unit factor.

7.3 The Equation $x^3 - 3y^3 = 1$

We apply the theory of the last section to show that $x^3 - 3y^3 = 1$ has no solutions in integers except for $(x, y) = (1, 0)$. Writing the equation in the form $x^3 - 1 = 3y^3$, we can factor the left side over I and consider it as

$$(x-1)(x-\omega)(x-\omega^2) = -\omega^2(1-\omega)^2 y^3 = -\omega^2(1-\omega)^{3m+2} z^3,$$

where $y = (1-\omega)^m z$ for some nonnegative integer m, and z is in I and not divisible by $1 - \omega$.

Exercise 3.1.

(a) Verify that the difference of any two of $x - 1$, $x - \omega$, and $x - \omega^2$ is the product of $1 - \omega$ and a unit.

(b) Prove that $1 - \omega$ must divide at least one of the factors $x - 1$, $x - \omega$, and $x - \omega^2$, and so it must divide each of the factors.

(c) Prove that a greatest common divisor of any pair of $x - 1$, $x - \omega$, and $x - \omega^2$ is $1 - \omega$.

(d) Prove that $(1-\omega)^2$ cannot divide more than one of $x - 1$, $x - \omega$, and $x - \omega^2$.

(e) Prove that $1 - \omega$ must divide y^3, so that $m > 0$.

(f) Prove that $x - 1$, $x - \omega$, and $x - \omega^2$ in some order have the forms $\epsilon_1(1-\omega)\gamma_1^3$, $\epsilon_2(1-\omega)\gamma_2^3$, and $\epsilon_3(1-\omega)^{3m}\gamma_3^3$ for units ϵ_i and numbers γ_i in I for which z is a unit times $\gamma_1\gamma_2\gamma_3$ and where each γ_i is not divisible by $1 - \omega$.

Exercise 3.2.

(a) Verify that $(x-1) + \omega(x-\omega) + \omega^2(x-\omega^2) = 0$ and deduce that

$$\gamma_1^3 + \epsilon\gamma_2^3 = \zeta(1-\omega)^{3m-1}\gamma_3^3 = 3\eta[(1-\omega)^{m-1}\gamma_3]^3$$

for some units ϵ, ζ, and η, where γ_1, γ_2, γ_3 are the quantities of Exercise 3.1(f).

(b) Prove that γ_1 and γ_2 are each congruent to ± 1, modulo $(1 - \omega)$, and deduce that for some choice of signs, $\pm 1 \pm \epsilon \equiv 0 \pmod{(1-\omega)^2}$.

(c) Check the possibilities ± 1, $\pm\omega$, $\pm\omega^2$ of units and conclude that $\epsilon \equiv \pm 1$. ♠

By relabeling γ_2 so that the minus sign is absorbed if necessary, we may assume that

$$\gamma_1^3 + \gamma_2^3 = 3\eta[(1-\omega)^{m-1}\gamma_3]^3.$$

Exercise 3.3.

(a) By factoring $\gamma_1^3 + \gamma_2^3 = (\gamma_1 + \gamma_2)(\gamma_1 + \omega\gamma_2)(\gamma_1 + \omega^2\gamma_2)$, imitate the argument in Exercise 3.1 to show that $m > 1$.

(b) By iterating the process that takes us from $x^3 - 1 = 3y^3$ to $\gamma_1^3 + \gamma_2^3 = 3\eta[(1-\omega)^{m-1}\gamma_3]^3$, obtain by descent a succession of equations of the latter type involving lower positive powers of $1 - \omega$ on the right. Deduce that the assumption of a solution in I for $x^3+1=3y^3$ must be false and that therefore $x^3-3y^3=1$ has no nontrivial solution in ordinary integers.

Exploration 7.1. Extend the method of this section to prove that $x^3+y^3=z^3$ cannot be solved in I and so the equation $(u+1)^3=u^3+v^3$ in Exercise 1.8 has no solution with $(u, v) \neq (-1, 1), (0, 1)$. Does $x^3-2y^3=1$ have a solution with $xy \neq 0$?

7.4 Obtaining the Cubic Version of Pell's Equation

Let c be any integer that is not a perfect cube, and let θ be its real cube root. In Chapter 2 we noted that the quadratic Pell's equation could be written in terms of a norm function involving the square root of d. We can proceed the same way for the cubic case. The number θ is the real root of the cubic equation

$$t^3-c=0.$$

This equation has three roots, namely θ, $\theta\omega$, and $\theta\omega^2$, where ω is the imaginary cube root of unity, $\frac{1}{2}(-1+i\sqrt{3})$.

Consider the expression $x+y\theta+z\theta^2$, where x, y, and z are integers. We define its *norm* by

$$N\big(x+y\theta+z\theta^2\big)=\big(x+y\theta+z\theta^2\big)\big(x+y\theta\omega+z(\theta\omega)^2\big)\big(x+y\theta\omega^2+z\big(\theta\omega^2\big)^2\big).$$

This will turn out to be a homogeneous polynomial of degree three in x, y, and z with integer coefficients. The analogue of Pell's equation will therefore be

$$N\big(x+y\theta+z\theta^2\big)=k.$$

Exercise 4.1. Noting that $\theta^3=c$, $\omega^2+\omega+1=0$, and $\omega^3=1$, verify that $N(x+y\theta+z\theta^2)=(x+y\theta+z\theta^2)[(x^2-cyz)+(cz^2-xy)\theta+(y^2-xz)\theta^2]=x^3+cy^3+c^2z^3-3cxyz$.

Exercise 4.2. Verify that $N((x+y\theta+z\theta^2)(u+v\theta+w\theta^2))=N(x+y\theta+z\theta^2)\cdot N(u+v\theta+w\theta^2)$.

Exercise 4.3. Suppose that $(x, y, z)=(u_1, v_1, w_1)$ is a solution of $x^3+cy^3+c^2z^3-3cxyz=1$. For each positive integer n, we can expand $(u_1+v_1\theta+w_1\theta^2)^n$ in the form $u_n+v_n\theta+w_n\theta^2$, by making the reduction $\theta^3=c$. From Exercise 4.2, argue that (u_n, v_n, w_n) is also a solution of $x^3+cy^3+c^2z^3-3cxyz=1$.

Exercise 4.4. Observe that $(x, y, z)=(1, 1, 1)$ is a solution of the equation $x^3+2y^3+4z^3-6xyz=1$. Use Exercise 4.3 to derive other solutions of this equation in positive integers. Check these.

Exercise 4.5.

(a) Verify the factorization

$$a^3 + b^3 + c^3 - 3abc = (a + b + c)(a^2 + b^2 + c^2 - ab - ac - bc).$$

(b) Use (a) to obtain the factorization

$$\begin{aligned} &x^3 + cy^3 + c^2z^3 - 3cxyz \\ &\quad = \frac{1}{2}(x + y\theta + z\theta^2)\big((x - y\theta)^2 + (y\theta - z\theta^2)^2 + (x - z\theta^2)^2\big). \end{aligned}$$

(c) Deduce from (b) that if (x, y, z) is a triple of large positive integers that satisfy the equation $x^3 + cy^3 + c^2z^3 - 3cxyz = 1$, then $x - y\theta$ and $y - z\theta$ must be close to zero, so that x/y and y/z are approximations of θ.

Exercise 4.6. Let x, y, and z be integers. Use the factorization of Exercise 4.5(b) to show that $(x + y\theta + z\theta^2)^{-1}$ has the form $(p + q\theta + r\theta^2)/K$, where p, q, r, K are all integers. Indeed, verify that

$$\begin{aligned} K &= x^3 + cy^3 + c^2z^3 - 3cxyz, \\ p &= x^2 - cyz, \\ q &= cz^2 - xy, \\ r &= y^2 - xz. \end{aligned}$$

Exercise 4.7. Note that if (u, v, w) is a solution of $x^3 + cy^3 + c^2z^3 - 3cxyz = 1$, then other solutions can be found from the expansion of negative integer powers of $u + v\theta + w\theta^2$. Use this to find solutions of $x^3 + 2y^3 + 4y^3 - 6xyz = 1$ in integers, not all of which are positive.

Exercise 4.8. So far, we have assumed that c is not a perfect cube. In this exercise we will see that when $c = a^3$ for some integer a, the behavior is quite different.

(a) Verify that

$$\begin{aligned} &2(x^3 + a^3y^3 + a^6z^3 - 3a^3xyz) \\ &\quad = (x + ay + a^2z)\big[(x - ay)^2 + a^2(y - az)^2 + (a^2z - x)^2\big]. \end{aligned}$$

(b) Suppose that $|a| \geq 2$ and that the integer triple (x, y, z) satisfies $x^3 + cy^3 + c^2z^3 - 3cxyz = 1$. Prove that $y = az$ and deduce that $x = a^2z \pm 1$. What do you conclude about the set of solutions in this case?

(c) Analyze the case $|a| = 1$.

7.5 Units

Let $\mathbf{Q}(\theta)$ be the set of real numbers of the form $u + v\theta + w\theta^2$, where u, v, w are rational numbers and θ^3 is the integer c; $\mathbf{Z}(\theta)$ is the subset of $\mathbf{Q}(\theta)$ for which

u, v, w are integers. An element ϵ of $\mathbf{Z}(\theta)$ is called a *unit* if $|N(\theta)| = 1$. Since $u^3 + cv^3 + c^2w^3 - 3cuvw = 1$ if and only if $u + v\theta + w\theta^2$ is a unit, we begin our study of the cubic Pell's equation by looking at the structure of the units in $\mathbf{Z}(\theta)$. The treatment is similar to that of the quadratic case in Section 4.1.

Exercise 5.1.

(a) Verify that $\mathbf{Q}(\theta)$ is a *field*; that is, the sum, difference, product, and quotient (with nonzero denominator) of two numbers in $\mathbf{Q}(\theta)$ also belong to $\mathbf{Q}(\theta)$.

(b) Verify that $\mathbf{Z}(\theta)$ is a *ring*; that is, the sum, difference, and product of two numbers in $\mathbf{Z}(\theta)$ also belong to $\mathbf{Z}(\theta)$.

(c) Prove that if $\alpha \in \mathbf{Z}(\theta)$, then also $N(\alpha)/\alpha \in \mathbf{Z}(\theta)$.

(d) Show that an element $\epsilon \in \mathbf{Z}(\theta)$ is a unit if and only if $1/\epsilon \in \mathbf{Z}(\theta)$. (Cf. the definition of unit in Section 7.2.)

(e) Show that if ϵ and η are units, then so is $\epsilon\eta$.

Exercise 5.2.

(a) Let $u + v\theta + w\theta^2 \in \mathbf{Q}(\theta)$. Define

$$\tau_1(u + v\theta + w\theta^2) = u + v\omega\theta + w\omega^2\theta^2,$$

$$\tau_2(u + v\theta + w\theta^2) = u + v\omega^2\theta + w\omega\theta^2,$$

where ω is an imaginary cube root of 1. Prove that for $\alpha, \beta \in \mathbf{Q}(\theta)$ and $i = 1, 2$;

$$\tau_i(\alpha \pm \beta) = \tau_i(\alpha) \pm \tau_i(\beta), \quad \tau_i(\alpha\beta) = \tau_i(\alpha)\tau_i(\beta), \quad \tau_i(1/\alpha) = 1/\tau_i(\alpha).$$

(These equations specify that τ_1 and τ_2 are *isomorphisms* of $\mathbf{Q}(\theta)$ into the field of complex numbers. The norm of an element α is the product $\alpha \cdot \tau_1\alpha \cdot \tau_2\alpha$.)

(b) Verify that $\tau_1(\alpha)$ is the complex conjugate of $\tau_2(\alpha)$ for $\alpha \in \mathbf{Q}(\theta)$.

(c) Verify that if $\alpha = u + v\theta + w\theta^2$, then α, $\tau_1(\alpha)$, and $\tau_2(\alpha)$ are the three roots of the cubic equation

$$t^2 - 3ut^2 + 3(u^2 - cvw)t - (u^3 + cv^3 + c^2w^3 - 3cuvw) = 0$$

with rational coefficients. Observe that if $\alpha \in \mathbf{Z}(\theta)$, then the cubic polynomial has integer coefficients.

Exercise 5.3. Determine all units $\epsilon \in \mathbf{Z}(\theta)$ whose absolute values $|\epsilon|$ are equal to 1.

Exercise 5.4. Let E be the set of units in $\mathbf{Z}(\theta)$ and suppose that E contains elements other than 1 and -1.

(a) Let M be a positive number exceeding 1. Prove that there are at most finitely many elements ϵ of E for which $1 \leq |\epsilon| \leq M$.

(b) Prove that E contains a smallest element γ that exceeds 1.

(c) Let $\epsilon \in E$, $|\epsilon| \neq 1$. Suppose that δ is the element among ϵ, $-\epsilon$, $1/\epsilon$, and $-1/\epsilon$ that exceeds 1. Prove that $\delta = \gamma^m$ for some positive integer m. Deduce that $\epsilon = \pm\gamma^n$ for some integer n. ♠

We turn to the question of the existence of a nontrivial unit whenever c is not a cube. The basic approach is similar to that used for the quadratic case in Section 4.2. From Exercise 4.1, we recall that

$$N(x + y\theta + z\theta^2) = (x + y\theta + z\theta^2)\big[(x^2 - cyz) + (cz^2 - xy)\theta + (y^2 - xz)\theta^2\big].$$

The strategy is to first show that for some real number M, $N(x + y\theta + z\theta^2) \le M$ occurs for infinitely many triples (x, y, z), so that $N(x + y\theta + z\theta^2)$ must assume some value infinitely often.

Exercise 5.5. Let n be an arbitrary positive integer and let the indices i and j satisfy $-n \le i, j \le n$.

(a) Explain why for each of the $(2n + 1)^2$ possible choices of the pair i, j we can select an integer a_{ij} for which $0 \le a_{ij} + i\theta + j\theta^2 < 1$.

(b) Use the pigeonhole principle to argue that for some positive integer k not exceeding $4n^2$, there are two distinct pairs of indices (i, j) for which the corresponding numbers $a_{ij} + i\theta + j\theta^2$ fall in the same interval

$$\left\{t : \frac{k-1}{4n^2} \le t \le \frac{k}{4n^2}\right\}.$$

(c) Deduce from (b) that there are integers u, v, w, not all zero, for which $|v| \le 2n$, $|w| \le 2n$, and $|u + v\theta + w\theta^2| \le 1/4n^2 \le 1/k(v, w)^2$, where $k(v, w) = \max(|v|, |w|)$. (Note that $k(v, w) \ge 1$.)

(d) Use the fact that

$$u + v\omega\theta + w\omega^2\theta^2 = (u + v\theta + w\theta^2) + v(\omega - 1)\theta + w(\omega^2 - 1)\theta^2$$

to show that

$$\left|u + v\omega\theta + w\omega^2\theta^2\right| \le \frac{1}{k(v, w)^2} + 4k(v, w)|c| \le 5k(v, w)|c|$$

and prove that

$$\left|N\left(u + v\theta + w\theta^2\right)\right| \le 25c^2.$$

(e) Prove that there are infinitely many triples (u, v, w) of integers for which $\left|N\left(u + v\theta + w\theta^2\right)\right| \le 25c^2$.

(f) Prove that there exists a positive integer m for which $N(x + y\theta + z\theta^2) = m$ has infinitely many solutions, with x, y, z integers.

Exercise 5.6. Let m be the positive integer found in Exercise 5.5.

(a) Prove that there are two distinct triples (u_1, v_1, w_1) and (u_2, v_2, w_2) of integers such that

$$N(u_1 + v_1\theta + w_1\theta^2) = N(u_2 + v_2\theta + w_2\theta^2) = m$$

and

$$u_1 \equiv u_2, \quad v_1 \equiv v_2, \quad w_1 \equiv w_2,$$

modulo m.

(b) Suppose that

$$m\left[\frac{u_1 + v_1\theta + w_1\theta^2}{u_2 + v_2\theta + w_2\theta^2}\right] = u_3 + v_3\theta + w_3\theta^2.$$

Prove that u_3, v_3, w_3 are integers each divisible by m.

(c) Let $u = u_3/m$, $v = v_3/m$, and $w = w_3/m$. Verify that (u, v, w) is a triple of integers distinct from $(1, 0, 0)$ for which $N(u + v\theta + w\theta^2) = 1$. This establishes that $x^3 + cy^3 + c^2z^3 - 3cxyz = 1$ is always solvable nontrivially for integers when the parameter c is not a cube.

7.6 Matrix and Vector Considerations

As for the quadratic Pell's situation, we induce from the multiplication of $u + v\theta + w\theta^2$ and $x + y\theta + z\theta^2$ a corresponding $*$-multiplication for triples (u, v, w) and (x, y, z) by

$$(u, v, w) * (x, y, z) = (ux + cvz + cwy, uy + vx + cwz, uz + vy + wx).$$

Let $(u, v, w)^{-1}$ be the triplet that corresponds to $(u + v\theta + w\theta^2)^{-1}$. If we think of (u, v, w) as being a fixed multiplier, we can describe its effect in matrix–vector form by

$$M\begin{pmatrix} x \\ y \\ z \end{pmatrix} = \begin{pmatrix} ux + cwy + cvz \\ vx + uy + cwz \\ wx + vy + uz \end{pmatrix},$$

where M is the 3×3 matrix

$$\begin{pmatrix} u & cw & cv \\ v & u & cw \\ w & v & u \end{pmatrix}.$$

For n an integer, let $(u, v, w)^n = (u_n, v_n, w_n)$ if $(u + v\theta + w\theta^2)^n = u_n + v_n\theta + w_n\theta^2$.

Setting $g_c(u, v, w) = u^3 + cv^3 + c^2w^3 - 3cuvw$, we define $(u, v, w)^0 = (1, 0, 0)$,

$$(u, v, w)^{-1} = \left(\frac{u^2 - cvw}{g_c(u, v, w)}, \frac{cw^2 - uv}{g_c(u, v, w)}, \frac{v^2 - uw}{g_c(u, v, w)}\right)$$

(cf. Exercise 4.1), and

$$(u, v, w)^{-n} = [(u, v, w)^{-1}]^n.$$

Suppose now that $g_c(x, y, z) = 1$ has solutions other than $(x, y, z) = (1, 0, 0)$ and that $(x, y, z) = (u, v, w)$ is the solution for which $u + v\theta + w\theta^2$ has the smallest positive value exceeding 1 (the fundamental solution). Then by Exercise 5.4, the entire set of solutions is given by $(x, y, z) = (u_n, v_n, w_n) = (u, v, w)^n$, where n is an integer. As in the quadratic case, we can find recursions satisfied by each of the sequences $\{u_n\}$, $\{v_n\}$, and $\{w_n\}$.

The sum of two 3×3 matrices and the product of a number and a matrix are defined componentwise as was done for 2×2 matrices in Section 1.2. The product of two 3×3 matrices is given by

$$\begin{pmatrix} a_{11} & a_{12} & a_{13} \\ a_{21} & a_{22} & a_{23} \\ a_{31} & a_{32} & a_{33} \end{pmatrix} \begin{pmatrix} b_{11} & b_{12} & b_{13} \\ b_{21} & b_{22} & b_{23} \\ b_{31} & b_{32} & b_{33} \end{pmatrix}$$
$$= \begin{pmatrix} a_{11}b_{11}+a_{12}b_{21}+a_{13}b_{31} & a_{11}b_{12}+a_{12}b_{22}+a_{13}b_{32} & a_{11}b_{13}+a_{12}b_{23}+a_{13}b_{33} \\ a_{21}b_{11}+a_{22}b_{21}+a_{23}b_{31} & a_{21}b_{12}+a_{22}b_{22}+a_{23}b_{32} & a_{21}b_{13}+a_{22}b_{23}+a_{23}b_{33} \\ a_{31}b_{11}+a_{32}b_{21}+a_{33}b_{31} & a_{31}b_{12}+a_{32}b_{22}+a_{33}b_{32} & a_{31}b_{13}+a_{32}b_{23}+a_{33}b_{33} \end{pmatrix}$$

Note that in the exercises, M is the matrix defined above and $g_c(u, v, w) = 1$.

Exercise 6.1. Let

$$I = \begin{pmatrix} 1 & 0 & 0 \\ 0 & 1 & 0 \\ 0 & 0 & 1 \end{pmatrix}.$$

Verify that for any 3×3 matrix A, $AI = IA = A$.

Exercise 6.2. Verify that

$$u_{n+1} = uu_n + cwv_n + cvw_n,$$
$$v_{n+1} = vu_n + uv_n + cww_n,$$
$$w_{n+1} = wu_n + vv_n + uw_n.$$

Exercise 6.3.

(a) Define

$$M^{-1} = \begin{pmatrix} u^2 - cvw & cv^2 - cuw & c^2w^2 - cuv \\ cw^2 - uv & u^2 - cvw & cv^2 - cuw \\ v^2 - uw & cw^2 - uv & u^2 - cvw \end{pmatrix}.$$

Verify that this definition is appropriate in that $MM^{-1} = M^{-1}M = I$.

(b) Verify that

$$M^2 = \begin{pmatrix} u^2 + 2cvw & 2cuw + cv^2 & 2cuv + c^2w^2 \\ 2vu + cw^2 & 2cvw + u^2 & cv^2 + 2cuw \\ 2wu + v^2 & cw^2 + 2uv & 2cvw + u^2 \end{pmatrix}.$$

(c) Verify that

$$M^2 - 3uM + 3(u^2 - cvw)I - M^{-1} = 0$$

and deduce that

$$M^3 - 3uM^2 + 3(u^2 - cvw)M - I = 0.$$

(d) Verify that

$$M \begin{pmatrix} \theta^2 \\ \theta \\ 1 \end{pmatrix} = (u + v\theta + w\theta^2) \begin{pmatrix} \theta^2 \\ \theta \\ 1 \end{pmatrix}.$$

(e) Use (c) to deduce that the sequences $\{u_n\}$, $\{v_n\}$, and $\{w_n\}$ each satisfy the recursion

$$x_{n+3} = 3ux_{n+2} - 3(u^2 - cvw)x_{n+1} + x_n.$$

Exercise 6.4. In Exercise 4.4, the solution $(x, y, z) = (1, 1, 1)$ was given for $g_2(x, y, z) = 1$. Determine $(1, 1, 1)^{-1}$ and use the recursion in Exercise 6.3(e) to derive other solutions. Check these.

Exercise 6.5.

(a) Verify that $g_3(x, y, z) = 1$ can be rewritten as

$$(x^3 - 1) + 3y^2 + 9(z^2 - xy)z = 0.$$

(b) Deduce that for any solution of (a), $x \equiv 1$ and $y \equiv 0$ modulo 3. Use these facts to obtain a solution by inspection.

(c) Determine other solutions by taking *-powers and check that the sequence of solutions you get satisfies the recurion in Exercise 6.3(e).

7.7 Solutions for Special Cases of the Parameter c

As in our initial investigation of the quadratic Pell's equation, it is possible to find solutions for

$$g_c(x, y, z) \equiv x^3 + cy^3 + c^2z^3 - 3cxyz = 1 \tag{1}$$

quite readily for certain values of c. This section will examine some ways of doing this.

Exercise 7.1.

(a) One strategy for locating a solution is to try $x = 1$. Then y and z must satisfy

$$y^3 + cz^3 - 3yz = 0. \tag{2}$$

Verify that if (2) is to be satisfied, then

$$c = \frac{-y(y^2 - 3z)}{z^3} = \frac{y(3z - y^2)}{z^3}.$$

Use this fact to determine values of c for which (2) has solutions with $z = 1$ and for which (1) has a solution with $x = z = 1$.

(b) Determine values of c for which (1) has a solution with $x = 1, z = -1$.

(c) Determine values of c for which (1) has a solution with $x = 1, z = \pm 2, \pm 3$.

(d) Make a table of some of these solutions (x, y, z) along with the inverse solutions $(x, y, z)^{-1}$ as defined in Section 7.6.

Exercise 7.2. Suppose that $c = -d$. Prove that $(x, y, z) = (u, v, w)$ is a solution of $x^3 + cy^3 + c^2z^3 - 3cxyz = 1$ if and only if $(x, y, z) = (u, -v, w)$ is a solution of $x^3 + dy^3 + d^2z^3 - 3dxyz = 1$. Thus, in analyzing Pell's equation, we can get the whole story essentially by looking at positive values of c.

Exercise 7.3.

(a) Suppose $c = k^3 + r$ where k and r are integers. Let s satisfy $rs = 3k$. Verify that $(x, y, z) = (1, ks, -s)$ is a solution of (1).

(b) The formula in (a) will always generate a *rational* solution for (1). Verify that for $c = 29$, a solution is $(x, y, z) = (1, 27/2, -9/2)$.

(c) We can specialize to $k = rt$, $s = 3t$ in (a), so that $c = r^3t^3 + r$ and $(x, y, z) = (1, 3rt^2, -3t)$ is a solution. List all of the positive values less than 300 for which we can find a solution in this way along with the corresponding solutions and inverse solutions.

(d) Specialize to the case that r is a multiple of 3 and obtain solutions for further values of c.

(e) Are there any other positive values of c not exceeding 100 for which we may obtain solutions? Try letting k be other than an integer.

Exercise 7.4. Consider the equation

$$x^3 + cy^3 + c^2z^3 - 3cxyz = 8, \tag{3}$$

where $c = a^3 + 2a$.

(a) Verify that this equation is satisfied by $(x, y, z) = (2, 3a, -3)$.

(b) By considering $(2 + 3a\theta - 3\theta^2)^2$, deduce and check that

$$(x, y, z) = (4 - 18ac, 12a + 9c, -12 + 9a^2)$$

is a solution of

$$x^3 + cy^3 + c^2z^3 - 3cxyz = 64. \tag{4}$$

(c) Show that when $a = 2b$ is even, the values of x, y, z in (b) are divisible by 4. In this case, verify that $c = 4b(2b^2 + 1)$ and that

$$(x, y, z) = (1 - 36b^2(2b^2 + 1), 3b(6b^2 + 5), 3(3b^2 - 1))$$

satisfies (1).

(d) Determine a solution of (1) in positive integers when $c = 12, 72$, and 228.

Exercise 7.5.

(a) Suppose that $c = r^2$. Prove that $(x, y, z) = (u, v, w)$ satisfies

$$x^3 + cy^3 + c^2z^3 - 3cxyz = 1$$

if and only if $(x, y, z) = (u, rw, v)$ satisfies

$$x^3 + ry^3 + r^2z^3 - 3rxyz = 1.$$

(b) Verify that $(x, y, z) = (4, 3, 2)$ satisfies (1) when $c = 3$ and use this result to obtain a solution to (1) when $c = 9$.

(c) Exercise 7.1 gave a method of solving equation (1) for $c = 5$. Use this solution to generate a solution (x, y, z) for which y is divisible by 5. From this, deduce a solution to (1) for $c = 25$.

Exercise 7.6.

(a) Suppose that $c = r^3s$. Prove that $(x, y, z) = (u, v, w)$ is a solution to

$$x^3 + cy^3 + c^2z^3 - 3cxyz = 1$$

if and only if $(x, y, z) = (u, rv, r^2w)$ is a solution to

$$x^3 + sy^3 + s^2z^3 - 3sxyz = 1.$$

(b) Find a solution in integers to the equation

$$x^3 + 16y^3 + 256z^3 - 48xyz = 1.$$

Exercise 7.7. Consider solutions of (1) with $z = 0$. In this case, the equation simplifies to $x^3 + cy^3 = 1$. With reference to Section 7.1, determine values of c for which a solution of this type is available along with some solutions and their inverses.

Exercise 7.8. Investigate solutions of $x^3 + cy^3 + c^2z^3 - 3cxyz = 1$ in the special cases that $c = k^3 \pm 1$ and $c = k^3 \pm 3$, and see whether you can find solutions that depend algebraically on the parameter k.

Exercise 7.9.

(a) Verify that

$$g_c(x + cz, x + y, y + z) = (c + 1)g_c(x, y, z).$$

(b) Starting with the fact that $g_c(1, 0, 0) = 1$, use (a) to determine at least two solutions to $g_c(x, y, z) = 1$ in positive rationals. Are there any situations in which integer solutions can be found in this way?

(c) Is one of the two solutions found in (b) a power of the other?

Exploration 7.2. Are there any values of c for which $g_c(x, y, z) = 1$ does *not* have a solutions with $x = 1$?

7.8 A Procedure That Often, but Not Always, Works

As seen in Exercise 1.5(b), when x, y, z are large positive integers, then $x^3 + cy^3 + c^2z^3 - 3cxyz = 1$ implies that x is close to $y\theta$ and y is close to $z\theta$. This suggests that we generate triple of integers (x, y, z) with these properties and hope that some of them will give us a solution. We start with four triples for which the signs of $x^3 - cy^3$ and $y^3 - cz^3$ together cover all four possibilities. Let $p = \lfloor \theta \rfloor$, the largest integer whose cube is less than c; let $q = \lfloor p\theta \rfloor$ and $r = \lfloor (p+1)\theta \rfloor$.

We form a table that begins

(x, y, z)	$x^3 - cy^3$	$y^3 - cz^3$
$(q, p, 1)$	$-$	$-$
$(q+1, p, 1)$	$+$	$-$
$(r, p+1, 1)$	$-$	$+$
$(r+1, p+1, 1)$	$+$	$+$

From this seed, we proceed as follows: Suppose the final entry in the table so far is (u, v, w). Let (u', v', w') be that last of the previous entries for which

$$u^3 - cv^3 \quad \text{and} \quad u'^3 - cv'^3$$

have opposite signs and also

$$v^3 - cw^3 \quad \text{and} \quad v'^3 - cw'^3$$

have opposite signs. The next entry is $(u + u', v + v', w + w')$.

Thus, the fifth entry in the table will have $(x, y, z) = (q + r + 1, 2p + 1, 2)$. The hope is that adding triples with opposite signs will keep bringing $x^3 - cy^3$ and $y^3 - cz^3$ relatively close to zero and so, in due course, make $x^3 + cy^3 + c^2z^3 - 3cxyz = 1$. Surprisingly, this works quite often; more surprisingly, it does not work all the time.

Exercise 8.1. For $c = 2$, verify that the algorithm yields the following table:

(x, y, z)	$x^3 - 2y^3$	$y^3 - 2z^3$	$x^3 + 2y^3 + 4z^3 - 6xyz$
(1, 1, 1)	$-$	$-$	1
(2, 1, 1)	$+$	$-$	2
(2, 2, 1)	$-$	$+$	4
(3, 2, 1)	$+$	$+$	11
(4, 3, 2)	$+$	$+$	6
(5, 4, 3)	$-$	$+$	1
(7, 5, 4)	$+$	$-$	9
(12, 9, 7)	$+$	$+$	22
(13, 10, 8)	$+$	$-$	5

Continue the table to generate more solutions to $x^3 + 2y^3 + 4z^3 - 6xyz = 1$, but check that the algorithm does not pick up $(x, y, z) = (281, 223, 177)$.

Exercise 8.2. For $c = 5$, verify that the algorithm yields

(x, y, z)	$x^3 - 5y^3$	$y^3 - 5z^3$	$x^3 + 5y^3 + 25z^3 - 15xyz$
(1, 1, 1)	−	−	16
(2, 1, 1)	+	−	2
(3, 2, 1)	−	+	8
(4, 2, 1)	+	+	9
(5, 3, 2)	−	−	10
(9, 5, 3)	+	−	4

Continue this table until a solution to $x^3 + 5y^3 + 25z^3 - 15xyz = 1$ is found.

Exercise 8.3. Try the algorithm to obtain solutions to

$$x^3 + cy^3 + c^2z^3 - 3cxyz = 1$$

when $c = 3, 4, 6, 7, 9, 10, 11, 12, 13, 14$. For the cases $c = 6, 10, 11, 13$, a pocket calculator is especially useful, and for the case $c = 12$, a programmable calculator or computer is desirable.

Exercise 8.4.

(a) The smallest positive solution of

$$x^3 + 15y^3 + 225z^3 - 45xyz = 1$$

is $(x, y, z) = (5401, 2190, 888)$. Check this solution and verify that the algorithm fails to find it.

(b) The smallest positive soution of

$$x^3 + 16y^3 + 256z^3 - 48xyz = 1$$

is $(x, y, z) = (16001, 6350, 2520)$. Check this solution and verify that the algorithm fails to find it.

Exercise 8.5. If you have suitable computational power at your disposal, check the efficacy of the algorithm for higher values of c.

7.9 A More General Cubic Version of Pell's Equation

So far, we have examined Pell's equation in the form $N(x + y\theta + z\theta^2) = 1$, where θ is a root of the special equation $t^2 - c = 0$. We extend our investigation to equations derived from roots of the cubic equation

$$t^3 + at^2 + bt + c = 0,$$

where a, b, and c are arbitrary integers. Suppose that the cubic polynomial cannot be factored as a product of polynomials of lower degree with integer coefficients and that its roots are $\theta = \theta_1, \theta_2$, and θ_3 with θ real. Define

$$g(x, y, z) = (x + y\theta_1 + z\theta_1^2)(x + y\theta_2 + z\theta_2^2)(x + y\theta_3 + z\theta_3^2).$$

The analogue of Pell's equation is $g(x, y, z) = 1$.

Exercise 9.1. Recall that $a = -(\theta_1 + \theta_2 + \theta_3)$, $b = \theta_1\theta_2 + \theta_1\theta_3 + \theta_2\theta_3$, and $c = -\theta_1\theta_2\theta_3$. Show that

$$g(x, y, z) = x^3 - cy^3 + c^2z^3 - ax^2y + (a^2 - 2b)x^2z + bxy^2 + acy^2z + (b^2 - 2ac)xz^2 - bcyz^2 + (3c - ab)xyz.$$

Exercise 9.2. Let θ be a root of the equation $t^3 = t + 1$.

(a) Verify that

$$g(x, y, z) = x^3 + y^3 + z^3 + 2x^2z + xz^2 - xy^2 - yz^2 - 3xyz.$$

(b) It turns out to be uncommonly easy to find solutions of $g(x, y, z) = 1$. By inspection, see how many you can get.

(c) We can obtain all solutions as $*$-powers of a fixed one (u, v, w), where $u + v\theta + w\theta^2$ has the smallest value exceeding 1. Do this.

(d) Using matrix techniques, determine a recursion satisfied by the sequence of solutions.

Exercise 9.3. Let θ be a root of the equation $t^3 - 7t^2 + 14t - 7 = 0$.

(a) Verify that $g(x, y, z) = x^3 + 7y^3 + 49z^3 + 7x^2y + 21x^2z + 14xy^2 + 49y^2z + 98xz^2 + 98yz^2 + 77xyz$.

(b) Determine some solutions of $g(x, y, z) = 1$ with $z = 0$.

(c) Determine some solutions of $g(x, y, z) = 1$ with $x = z = 1$.

(d) List other solutions. Do you think that they are all obtainable as $*$-powers of a single solution?

7.10 More Explorations

Exploration 7.3. Consider the function $g_2(x, y, z) = x^3 + 2y^3 + 4z^3 - 6xyz$. There appear to be a number of interesting regularities that occur, as, for example, in the following table:

n	(x, y, z)	$g_2(x, y, z)$
0	(1, 0, 0)	1
1	(1, 1, 0)	3
2	(1, 1, 1)	1
3	(3, 2, 2)	3
4	(5, 4, 3)	1
5	(11, 9, 7)	3

If (x_n, y_n, z_n) is the nth triple, then $\{x_n\}$, $\{y_n\}$, and $\{z_n\}$ each appear to satisfy the recursion

$$t_{2m} = t_{2m-1} + t_{2m-2} + t_{2m-3},$$

$$t_{2m+1} = 2t_{2m} + t_{2m-2},$$

for $m \geq 2$.

Another table of regularities is

n	(x, y, z)	$g_2(x, y, z)$
0	$(1, 0, 0)$	1
1	$(0, 1, 0)$	2
2	$(1, -1, 1)$	9
3	$(0, 0, 1)$	4
4	$(1, 0, 1)$	5
5	$(1, 1, 1)$	1
6	$(2, 1, 1)$	2
7	$(1, 2, 1)$	9
8	$(2, 2, 1)$	4
9	$(3, 3, 2)$	5
10	$(5, 4, 3)$	1

In this case, the recursion seems to be, for $m \geq 1$,

$$\begin{aligned} t_{5m-1} &= t_{5m-3} + t_{5m-4}, \\ t_{5m} &= t_{5m-1} + t_{5m-4}, \\ t_{5m+1} &= t_{5m} + t_{5m-5}, \\ t_{5m+2} &= t_{5m} + t_{5m-4}, \\ t_{5m+3} &= t_{5m+1} + t_{5m-4}. \end{aligned}$$

It seems to happen more frequently than one would expect that

$$g_2(x_1, y_1, z_1) + g_2(x_2, y_2, z_2) = g_2(x_1 + x_2, y_1 + y_2, z_1 + z_2).$$

For example,

$$\begin{aligned} &g_2(1, 1, 1) + g_2(5, 4, 3) = g_2(6, 5, 4), \\ &g_2(5, 4, 3) + g_2(8, 6, 5) = g_2(13, 10, 8), \\ &g_2(1, 1, 0) + g_2(1, 1, 1) = g_2(2, 2, 1), \\ &g_2(1, 0, 0) + g_2(3, 3, 2) = g_2(4, 3, 2). \end{aligned}$$

Can anything be said in general?

Exploration 7.4. Let c be a noncubic integer and θ its real cube root. The number $x + y\theta + z\theta^2$ in $\mathbf{Q}(\theta)$ is an algebraic integer if and only if it is a root of a monic polynomial with integer coefficients. The monic cubic polynomial whose roots are $x + y\theta + z\theta^2$, $x + y\omega\theta + z\omega^2\theta^2$, $x + y\omega^2\theta + z\omega\theta^2$ is $t^3 - pt^2 + qt - r$ where $p = 3x$, $q = 3(x^2 - cyz)$, and $r = x^3 + cy^3 + c^2z^3 - 3cxyz$. Now, p, q, r are certainly integers when x, y, z are themselves integers. Are there values of c for which algebraic integers exist where x, y, z are not all (ordinary) integers, but p, q, r are integers?

7.11 Notes

Section 3. For an account of $\mathbf{Q}(\sqrt{-3})$, consult G.H. Hardy and E.M. Wright, *An Introduction to the Theory of Numbers* (Oxford), Chapter XIII. In the fourth edition (1960), the relevant material is found on pages 188–189 and 192–196. This chapter gives a treatment of a few special cases of Fermat's "theorem" that there are no nontrivial solutions in integers of $x^n + y^n = z^n$ when n is a positive integer exceeding 2. This is the theorem that was finally settled by Andrew Wiles in the last decade of the twentieth century.

5.5–5.6. See G.B. Mathews, *On the Arithmetic Theory of the Form* $x^3 + ny^3 + n^2z^3 - 3nxyz$, *Proc. London Mathematical Society* 21 (1890), 280–287.

7.9. When x, y, z are positive and $g_c(x, y, z) = 1$, then y/z, x/y, and cz/x are approximations to $c^{1/3}$ whose product is c. Thus some are over- and others under-approximations. We can add numerators and denominators to get better approximations $(x+y)/(y+z)$, $(x+cz)/(y+x)$. We select the third approximation to make the product of the three to be equal to c: $c(y+z)/(x+cz)$. This motivates the transformation

$$S(x, y, z) = (x + cz, y + x, z + y).$$

Compare the values of $g_c(x + cz, y + x, z + y)$, $g_c(cz, x, y)$, and $g_c(x, y, z)$.

The tranformation is related to the following algorithm for determining the cube root of any positive number c. Begin with the quadruple $(1, 1, 1, c)$. We form a sequence of quadruples in which (p, q, r, cp) is followed by $(p + q, q + r, r + cp, c(p+q))$. It turns out that as one proceeds along the sequence, $q/p, r/q, cp/r$ all approach $c^{1/3}$. This can be generalized to higher roots. Thus, for the kth root of c, start with $(1, 1, 1, \ldots, 1, c)$ (with k ones) and apply the transformation

$$(p, q, r, \ldots, s, cp) \longrightarrow (p + q, q + r, \ldots, s + cp, c(p + q)),$$

where each of the first k entries is the sum of the corresponding entry and its successor in the previous vector.

Section 9. A recent researcher who has done a significant amount of work on the determination of cubic fields is T.W. Cusick; an example of the work of him and his colleagues is listed in the bibliography.

7.12 Hints

2.2. Let $\phi(x + y\omega) = x + y\omega^2$, the surd conjugate of $x + y\omega$. Show that $\phi(r) = r$ for each rational r, $\phi(\alpha \pm \beta) = \phi(\alpha) \pm \phi(\beta)$, $\phi(\alpha\beta) = \phi(\alpha)\phi(\beta)$. Note that $N(\alpha) = \alpha\phi(\alpha)$.

2.10. Since $\sigma | \rho$, $\rho = \sigma\epsilon$ for some algebraic integer ϵ. Since $\rho | \sigma$, deduce that ϵ^{-1} must also be an algebraic integer.

2.12(b). Note that $\mu = \xi\alpha\mu + \eta\beta\mu$ and that β divides each term on the right side.

3.2(b). Let $\gamma_1 = r_1 + s_1\omega = (r_1 + s_1) - s_1(1 - \omega)$ and note that $r_1 + s_1$ is not divisible by $3 = (1 - \omega)(1 - \omega^2)$.

5.1(a). Use the fact that $\theta^3 = c$ to check products and refer to Exercise 1.7 to show that the reciprocal of an element in $\mathbf{Q}(\theta)$ is also in $\mathbf{Q}(\theta)$.

5.1(d). Observe that if $\alpha \in \mathbf{Z}(\theta)$, then $N(\alpha)$ must be an integer. Use the fact that $N(1/\alpha) = 1/N(\alpha)$.

5.2(c). Determine the coefficients by looking at the sum, sum of products of pairs, and products of the roots. Use the fact that $\omega^3 = 1$ and $\omega + \omega^2 = -1$.

5.4(a). Observe that $\tau_1(\epsilon)$ and $\tau_2(\epsilon)$ are complex conjugates and use the fact that $\epsilon\tau_1(\epsilon)\tau_2(\epsilon) = 1$ to determine bounds for $|\tau_1(\epsilon)|$ and $|\tau_2\epsilon|$. Now use the symmetric functions of ϵ, $\tau_1(\epsilon)$, and $\tau_2(\epsilon)$ to find bounds for the coefficients of ϵ (cf. Exercise 5.2(b)).

5.5(c). Take the difference of the numbers in the pair found in (b).

5.5(e). Select (u_1, v_1, w_1) such that $N(u_1 + v_1\theta + w_1\theta^2) \leq 25|c|^2$. Determine an integer n such that $1/4n^2 < |u_1 + v_1\theta + w_1\theta^2|$, and use (c) to find a distinct triple (u_2, v_2, w_2) with $N(u_2 + v_2\theta + w_2\theta \equiv 2) \leq 25c^2$. Continue on in this way, churning out an infinite sequence of triples.

5.6(a). Note that there are only finitely many equivalence classes, modulo m, available for the triple (u, v, w). Use the pigeonhole principle.

8

Analogues of the Fourth and Higher Degrees

In this chapter, we consider analogues of Pell's equation of higher degree, particularly the fourth. Throughout, n is a positive integer and c is an integer which has a real nth root θ of the same sign. Define

$$\begin{aligned} g_c(x_1, x_2, \ldots, x_n) &\equiv N(x_1 + x_2\theta + x_3\theta^2 + \cdots + x_n\theta^{n-1}) \\ &\equiv \prod_{i=0}^{n} \left(x_1 + x_2(\zeta^i\theta) + \cdots + x_n(\zeta^i\theta)^{n-1}\right) \end{aligned}$$

where $\zeta = \cos\frac{2\pi}{n} + i\sin\frac{2\pi}{n}$ is a primitive nth root of unity. The analogue of Pell's equation that we want to examine is

$$g_c(x_1, x_2, \ldots, x_n) = 1.$$

8.1 Solution of Some Special Cases for General n

After reviewing some basic properties of roots of unity, we consider the solutions to Pell's equation for $c = 2, 3$ and $k^n \pm 1$, where k is a positive integer.

Exercise 1.1.

(a) Verify that $1, \zeta, \zeta^2, \ldots, \zeta^{n-1}$ are distinct complex numbers that satisfy the equation $t^n = 1$.

(b) Deduce, from the factor theorem, that $t^n - 1 = (t-1)(t-\zeta)(t-\zeta^2)\cdots(t-\zeta^{n-1})$.

(c) Show that ζ^i $(1 \le i \le n-1)$ satisfies the equation $1+t+t^2+\cdots+t^{n-1} = 0$.

Exercise 1.2. Let c and θ be real numbers for which $\theta^n = c$.

(a) Prove that the nth roots of c are $\theta, \zeta\theta, \zeta^2\theta, \ldots, \zeta^{n-1}\theta$ and that

$$t^n - c = (t-\theta)(t-\zeta\theta)\cdots(t-\zeta^{n-1}\theta).$$

(b) When n is odd, prove that

$$t^n + c = (t+\theta)(t+\zeta\theta)\cdots(t+\zeta^{n-1}\theta).$$

Exercise 1.3.
(a) Verify that

$$u^{n-1} + u^{n-2}v + \cdots + uv^{n-2} + v^{n-1} = \frac{u^n - v^n}{u - v}.$$

(b) Deduce from (a) that with $\theta^n = c$,

$$k^{n-1} + k^{n-2}(\zeta^i\theta) + \cdots + k(\zeta^i\theta)^{n-2} + (\zeta^i\theta)^{n-1} = \frac{k^n - c}{k - \zeta^i\theta}.$$

Exercise 1.4. Prove that

$$g_c\left(k^{n-1}, k^{n-2}, \ldots, k, 1\right) = \prod_{i=0}^{n-1}\left(\frac{k^n - c}{k - \zeta^i\theta}\right) = (k^n - c)^{n-1}.$$

Exercise 1.5. Let $c = k^n - 1$.
(a) Prove that $N(k - \theta) = 1$ and so $g_c(k, -1, 0, 0, \ldots, 0) = 1$.
(b) Verify that (a) corroborates known results when $n = 2, 3$.
(c) Deduce from Exercise 1.4 that $(x_1, x_2, \ldots, x_n) = \left(k^{n-1}, k^{n-2}, \ldots, k, 1\right)$ is a solution of $g_c(x_1, x_2, \ldots, x_n) = 1$.

Exercise 1.6. Let $c = k^n + 1$.
(a) Prove that $N(k - \theta) = -1$, and so $g_c(k, -1, 0, 0, \ldots, 0) = -1$.
(b) Verify that (a) corroborates known results when $n = 2, 3$.
(c) Deduce from Exercise 1.4 that $(x_1, x_2, \ldots, x_n) = (k^{n-1}, k^{n-2}, \ldots, 1)$ is a solution of $g_c(x_1, x_2, \ldots, x_n) = (-1)^{n-1}$.

Exercise 1.7.
(a) Verify that $N(k + \theta) = k^n - (-1)^n c$.
(b) Deduce from (a) that when n is even, $g_{k^n-1}(k, 1, 0, \ldots, 0) = 1$ and $g_{k^n+1}(k, 1, 0, \ldots, 0) = -1$.

Exercise 1.8.
(a) Prove that

$$N(k^2 + k\theta + \theta^2) = \begin{cases} \dfrac{k^{3n} - c^3}{k^n - c}, & \text{when } n \text{ is not a multiple of 3,} \\ (k^n - c)^2, & \text{when } n \text{ is a multiple of 3.} \end{cases}$$

(b) If n is a multiple of 3 and $c = k^n \pm 1$, verify that

$$g_{k^n\pm 1}(k^2, k, 1, 0, \ldots, 0) = 1.$$

Exercise 1.9. If $\xi = x_1 + x_2\theta + \cdots + x_n\theta^{n-1}$ and $\eta = y_1 + y_2\theta + \cdots + y_n\theta^{n-1}$, prove that $N(\xi\eta) = N(\xi)N(\eta)$. Deduce that if $N(\xi) = -1$, then $N(\xi^2) = +1$.

Exercise 1.10.
(a) If $\xi = k^{n-1} + k^{n-2}\theta + \cdots + k\theta^{n-2} + \theta^{n-1}$, show that
$$\xi^2 = \left[k^{2n-2} + (n-1)ck^{n-2}\right] + \left[2k^{2n-3} + (n-2)ck^{n-3}\right]\theta + \left[3k^{2n-4} + (n-3)ck^{n-4}\right]\theta^2 + \cdots + nk^{n-1}\theta^{n-1}.$$

(b) Suppose that n is even and that $c = k^n + 1$ for some positive integer k. Explain how to find a solution to $g_c(x_1, x_2, \ldots, x_n)$.

Exercise 1.11. Suppose that n is even and $c = k^n \pm 2$. Prove that
$$g_c(k^n \pm 1, k^{n-1}, k^{n-2}, \ldots, k) = 1.$$

Exercise 1.12. Consider the case $c = 2$.
(a) Using the fact that $g_2(1, 1, 1, \ldots, 1) = (-1)^{n-1}$, show that $(x_1, x_2, \ldots, x_n) = (2n-1, 2n-2, \ldots, n+1, n)$ satisfies $g_2(x_1, x_2, \ldots, x_n) = 1$. Check directly from the definition of N that
$$N\big((2n-1) + (2n-2)\theta + \cdots + n\theta^{n-1}\big) = 1.$$

(b) Determine $N(3 + 2\theta + 2\theta^2 + \cdots + 2\theta^{n-1})$. For which values of n do we obtain a solution of $g_2(x_1, x_2, \ldots, x_n) = 1$?

Exercise 1.13.
(a) Verify that
$$g_3(2, 1, 1, \ldots, 1) = \prod_{i=0}^{n-1}(2 + (\zeta^i\theta) + \cdots + (\zeta^i\theta)^{n-1}) = (-1)^n \prod_{i=0}^{n-1} \frac{(1 + \zeta^i\theta)}{(1 - \zeta^i\theta)}.$$

(b) Suppose that n is even. Show that $\{-\zeta^i : i = 0, \ldots, n-1\}$ consists of all the nth roots of unity. Deduce from (a) that $g_3(2, 1, 1, \ldots, 1) = 1$.

Exercise 1.14. Verify that $g_3(3k-2, 3k-3, \ldots, k+1, k) = 1$ when $n = 2k-1$.

8.2 The Quartic Pell's Equation with Positive Parameter

The *norm form* of $x + y\theta + z\theta^2 + w\theta^3$, for $\theta^4 = c$, is equal to
$$g_c(x, y, z, w) = (x + y\theta + z\theta^2 + w\theta^3)(x - y\theta + z\theta^2 - w\theta^3) \times (x + iy\theta - z\theta^2 - iw\theta^3)(x - iy\theta - z\theta^2 + iw\theta^3).$$

As for the cases of the quadratic and cubic Pell's equations, we can induce a *-product for quadruples (x, y, z, w) from the ordinary multiplication of reals

$x + y\theta + z\theta^2 + w\theta^3$. The powers $(x, y, z, w)^n$ for integer n can be defined as before with $(x, y, z, w)^0 = (1, 0, 0, 0)$.

For most of this section we restrict attention to positive values of c. Negative values of c are considered in the next section.

Exercise 2.1. Verify that

$$g_c(x, y, z, w) = (x^2 + cz^2 - 2cyw)^2 - c(2xz - y^2 - cw^2)^2.$$

Exercise 2.2. Verify that

$$\begin{aligned}(x_1, y_1, z_1, w_1) &* (x_2, y_2, z_2, w_2)\\ &= (x_1x_2 + cy_1w_2 + cz_1z_2 + cw_1y_2, x_1y_2 + y_1x_2 + cz_1w_2 + cw_1z_2,\\ &\qquad x_1z_2 + y_1y_2 + z_1x_2 + cw_1w_2, x_1w_2 + y_1z_2 + z_1y_2 + w_1x_2)\end{aligned}$$

and that

$$\begin{aligned}g_c(x, y, z, w)(x, y, z, w)^{-1} = (&x^3 + cy^2z + c^2zw^2 - cxz^2 - 2cxyw,\\ &cy^2w + 2cxzw - x^2y - cyz^2 - c^2w^3,\\ &xy^2 + cxw^2 + cz^3 - x^2z - 2cyzw,\\ &cyw^2 + 2xyz - y^3 - x^2w - cz^2w).\end{aligned}$$

Exercise 2.3. Verify that if $g_c(x, y, z, w) = 1$, then

$$(x, y, z, w)^{-1} = (x, -y, z, -w) * (x^2 + cz^2 - 2cyw, 0, y^2 + cw^2 - 2xz, 0).$$

Exercise 2.4. Suppose that $c = b^4$ for some positive integer b. Does the equation $g_c(x, y, z, w) = 1$ have only finitely many solutions in this case? Explain.

Exercise 2.5. Suppose that $c = a^2$ for some positive nonsquare integer a. Then $\theta = a^{1/2}$, and $x + y\theta + z\theta^2 + w\theta^3$ collapses to $(x + az) + (y + aw)\theta$, where θ is now an irrational number satisfying a quadratic equation with integer coefficients. The quadratic norm form is $(x + az)^2 - (y + aw)^2a$. In this situation, the quartic Pell's equation $g_c(x, y, z, w) = 1$ will be of a different character from that when c is not a perfect square. However, we can still see what can be said about its solutions.

(a) Verify that in this case, the equation to be solved is

$$(x^2 + a^2z^2 - 2a^2yw)^2 - a^2(2xz - y^2 - a^2w^2)^2 = 1.$$

(b) Argue that the equation in (a) is equivalent to the system

$$\begin{aligned}x^2 + a^2z^2 - 2a^2yw &= \pm 1,\\ y^2 + a^2w^2 - 2xz &= 0.\end{aligned}$$

(c) Deduce from (b) that

$$(x - az)^2 + a(y - aw)^2 = \pm 1,$$
$$(x + az)^2 - a(y + aw)^2 = \pm 1,$$

and note that this implies that the sign on the right side must be $+$.

(d) Deduce from (c) that $y = aw$ and $x - az = \pm 1$.

Exercise 2.6. For the equation of Exercise 2.5, we consider the case that $x - az = 1$ and $y = aw$.

(a) Obtain the equation $a^2w^2 = z(1 + az)$.

(b) From (a), show that there are integers r and s for which $z = a^2s^2$ and $1 + az = r^2$, so that

$$r^2 - a^3s^2 = 1.$$

(c) Suppose that (r, s) satisfies $r^2 - a^3s^2 = 1$. Show that $(x, y, z, w) = (r^2, ars, a^2s^2, rs)$ is a solution of $g_{a^2}(x, y, z, w) = 1$.

(d) Use the foregoing to determine solutions to $g_c(x, y, z, w) = 1$, when $c = 4, 9, 25$.

Exercise 2.7. For the equation of Exercise 2.5, carry out an analysis similar to that in Exercise 2.6 for the possibility that $x - az = -1$ and $y = aw$, and thence determine a solution to $g_{25}(x, y, z, w) = 1$. ♠

Henceforth, we will assume that c is a positive integer that is not a perfect square. Let $p_c(x, y, z, w) = x^2 + cz^2 - 2cyw$ and $q_c(x, y, z, w) = 2xz - y^2 - cw^2$. Then $g_c(x, y, z, w) = p_c(x, y, z, w)^2 - cq_c(x, y, z, w)^2$, so that every solution of $g_c(x, y, z, w) = 1$ gives rise to a solution $(r, s) = (p_c, q_c)$ of the quadratic Pell's equation $r^2 - cs^2 = 1$. Some of these quartic solutions lead to the trivial $(r, s) = (\pm 1, 0)$ (we will call these Type A), while others lead to nontrivial quadratic solutions (Type B). What can be said about the relation between the two types of Pell's equation? We begin by getting some numerical information.

We try an algorithm similar to that used for the cubic case. Let θ be the positive fourth root of c. We begin with a "base" of eight quadruples (x, y, z, w) corresponding to the sign combinations $(\pm, \pm, \pm)$. In each case, $w = 1$. For each sign combination $(\pm, \pm, -)$, let $z = \lfloor \theta \rfloor$, i.e., the largest integer for which $z^4 - cw^4 < 0$; for $(\pm, \pm, +)$, let $z = \lceil \theta \rceil$, i.e., the smallest integer for which $z^4 - cw^4 > 0$.

We have indicated how w and z are to be found in each case. For $(\pm, -, \pm)$, y is the largest integer for which $y^4 - cz^4 < 0$, while for $(\pm, +, \pm)$, y is the smallest integer for which $y^4 - cz^4 > 0$. Finally, for $(-, \pm, \pm)$, x is the largest integer for which $x^4 - cy^4 < 0$, and for $(+, \pm, \pm)$, x is the smallest integer for which $x^4 - cy^4 > 0$.

Exercise 2.8.

(a) Verify that for $c = 8$, the base is given in the following table:

(x, y, z, w)	$(x^4 - 8y^4)$	$(y^4 - 8z^4)$	$(z^4 - 8w^4)$	sign	p_c	q_c	g_c
$(1, 1, 1, 1)$	-7	-7	-7	$(-, -, -)$	-7	-7	-343
$(2, 1, 1, 1)$	8	-7	-7	$(+, -, -)$	-4	-5	-184
$(3, 2, 1, 1)$	-47	8	-7	$(-, +, -)$	-15	-6	-63
$(4, 2, 1, 1)$	128	8	-7	$(+, +, -)$	-8	-4	-64
$(5, 3, 2, 1)$	-23	-47	8	$(-, -, +)$	9	3	9
$(6, 4, 2, 1)$	-752	128	8	$(-, +, +)$	4	0	16
$(6, 3, 2, 1)$	648	-47	8	$(+, -, +)$	20	7	8
$(7, 4, 2, 1)$	353	128	8	$(+, +, +)$	17	4	161

To continue the table, take any two vectors (x, y, z, w) with opposite sign combinations, say $(+, -, +)$, and $(-, +, -)$ and take the vector sum. This may yield only one vector; if not, construct a different table for each possibility. After this, obtain each subsequent entry by taking the vector sum of the last entry found and the previous entry of the opposite sign combination. Verify that the table can proceed with the next entry $(8, 5, 3, 2)$ $(-, +, -)$ or $(9, 5, 3, 2)$ $(+, +, -)$. Continue in this manner.

Exercise 2.9. From Exercise 2.8, we note that $g_8(6, 4, 2, 1) = 16$, $p_8(6, 4, 2, 1) = 4$, and $q_8(6, 4, 2, 1) = 0$. Use these equations to obtain a rational solution to Pell's equation $g_8(x, y, z, w) = 1$.

Exercise 2.10.

(a) Follow the procedure just described to obtain three solutions of $g_2(x, y, z, w) = \pm 1$. Investigate whether these solutions are related with respect to *-multiplication.

(b) Follow the procedure just described until it yields three solutions of $g_3(x, y, z, w) = 1$.

Exercise 2.11. Certain solutions of $g_c(x, y, z, w) = 1$ are easy to find. Verify that $g_c(u, 0, v, 0) = 1$ whenever $u^2 - cv^2 = \pm 1$. Which of these solutions are of Type B?

Exercise 2.12. Suppose that $u^2 - cv^2 = 1$. Then it is possible to construct a solution for $g_c(x, y, z, w) = 1$ with $(x, z) = (u, v)$ for certain values of c. Suppose r and s are selected so that $rs = v^2$.

(a) Verify that $p_c(u, r, v, s) = 1$.

(b) Prove that the condition $q_c(u, r, v, s) = 0$ is equivalent to $r^2 = v(u + 1)$ or $r^2 = v(u - 1)$.

(c) Investigate values of c for which a solution to the quartic equation can be obtained from a solution to the corresponding quadratic equation in the way outlined above.

Exercise 2.13. A special case of Exercise 2.12 is $c = (2k)^4 \pm 4 = 4(4k^4 \pm 1)$, where the parameter c differs from an even fourth power by 4. We recall that when $d = (2m)^2 \pm 4$, then $x^2 - dy^2 = 1$ is satisfied by $(x, y) = (2m^2 \pm 1, m)$. Apply this to the case $m = 2k^2$ to verify that $u^2 - cv^2 = 1$ holds with $(u, v) = (8k^4 \pm 1, 2k^2)$. Use the procedure of Exercise 2.12 to obtain solutions for $g_c(x, y, z, w) = 1$ with $xyzw \neq 0$.

Exercise 2.14. Suppose that $y^2 + cw^2 - 2xz = 0$.

(a) Verify that

$$x^2 + cz^2 - 2cyw = \left(x - z\sqrt{c}\right)^2 + \sqrt{c}\left(y - w\sqrt{c}\right)^2.$$

Deduce that for any Type A solution, $x^2 + cz^2 - 2cyw = -1$ cannot occur.

(b) Here is an alternative way of establishing the result in (a). Assume that Type A solutions with $x^2 + cz^2 - 2cyw = -1$ can occur. When c is odd, observe that y and w must have the same parity, while x and z must have opposite parity. If $c \equiv 1 \pmod 4$ with y and w odd, show that xz is also odd and get a contradiction.

On the other hand, if $c \equiv 1 \pmod 4$ with y and w even, use the fact that $x^2 + cz^2 - 2cyw \equiv x^2 + z^2 \pmod 4$ to get a contradiction. If $c \equiv 3 \pmod 4$, prove that c must be divisible by a prime p congruent to 3 modulo 4; use the fact that for such a prime, $x^2 \equiv -1 \pmod p$ is not solvable. Next, show that we get a contradiction when c is divisible by 4. Finally, deal with the cases $c \equiv 2$ and $c \equiv 6 \pmod 8$.

Exercise 2.15.

(a) Suppose that $x^2 + cz^2 - 2cyw = p$ and $2xz - y^2 - cw^2 = q$, where $p^2 - cq^2 = 1$. Using the fact that

$$p - q\sqrt{c} = \left(x - z\sqrt{c}\right)^2 + \sqrt{c}\left(y - w\sqrt{c}\right)^2,$$

deduce that $p - q\sqrt{c} > 0$ and thence that $p > 0$.

(b) Deduce from (a) that for $c = 3, 6, 8$, $g_c(x, y, z, w) = 1$ has no Type B solution corresponding to a fundamental solution $(x, y) = (p, |q|)$ of the corresponding quadratic pellian equation $x^2 - cy^2 = 1$.

Exercise 2.16. Prove that $g_c(x, y, z, w) = 1$ has a Type B solution corresponding to a fundamental solution of $x^2 - cy^2 = 1$ whenever $x^2 - cy^2 = -1$ is solvable in integers.

Exercise 2.17. Here is an alternative way of showing that $g_3(x, y, z, w) = 1$ has no Type B solution corresponding to the solution $(\pm 2, \pm 1)$ of $x^2 - 3y^2 = 1$. Suppose that

$$|x^2 + 3z^2 - 6yw| = 2,$$
$$|y^2 + 3w^2 - 2xz| = 1.$$

Deduce that y and w must have opposite parity, and so $x^2 + 3z^2 \equiv 2 \pmod 4$. Obtain a contradiction.

Exercise 2.18. If we specialize to solutions for which $z = w = 0$, the equation $g_c(x, y, z, w) = 1$ becomes $x^4 - cy^4 = 1$. This will not have a solution for every parameter c. However, there are generic situations in which solutions can be found. Let d be a (not necessarily positive) integer and suppose that either $u^2 - dv^2 = 1$ or $u^2 + dv^2 = -1$, and that $2d$ is a multiple of v^2.

(a) Verify that

$$u^4 - \left(d^2 + \frac{2d}{v^2}\right)v^4 = 1.$$

(b) Use this to determine values of c for which $x^4 - cy^4 = 1$ is solvable and display the solution.

Exercise 2.19. In this exercise we investigate the structural relationship between the set of solutions to the quartic Pell's equation $g_c(x, y, z, w) = 1$ and the corresponding quadratic Pell's equation. Let

$$\mathbf{Z}(\theta) \equiv \{(x + y\theta + z\theta^2 + w\theta^3 : x, y, z, w \in \mathbf{Z}\}$$

and

$$\mathbf{Z}(\sqrt{c}) \equiv \{p + q\sqrt{c} : p, q \in \mathbf{Z}\}.$$

Each of these is a set of real numbers closed under the arithmetic operations of addition, subtraction, and multiplication of pairs of its elements. We can think of each set as generalizing the set of ordinary integers.

Recall that the mappings

$$\lambda(x + y\theta + z\theta^2 + w\theta^3) = (x, y, z, w)$$

and

$$\mu(p + q\sqrt{c}) = (p, q)$$

are one-to-one mappings from the sets of elements of norm 1 in $\mathbf{Z}(\theta)$ and $\mathbf{Z}(\sqrt{c})$, respectively, to solutions of the corresponding Pell's equation that satisfy

$$\lambda(\alpha\beta) = \lambda(\alpha) * \lambda(\beta)$$

and

$$\mu(\gamma\delta) = \mu(\gamma) * \mu(\delta).$$

(a) Verify that the mapping

$$\rho(x + y\theta + z\theta^2 + w\theta^3) = (x^2 + cz^2 - 2cyw) + \sqrt{c}(y^2 + cw^2 - 2xz)$$

defined on the set of elements of norm 1 in $\mathbf{Z}(\theta)$ to the elements of norm 1 in $\mathbf{Z}(\sqrt{c})$ also satisfies

$$\rho(\alpha\beta) = \rho(\alpha)\rho(\beta).$$

(b) Let $\sigma = \mu \circ \rho \circ \lambda^{-1}$. Prove that

$$\sigma(\alpha * \beta) = \sigma(\alpha) * \sigma(\beta)$$

(so that σ is a *homomorphism* from the group of solutions of $g_c(x, y, z, w) = 1$ to the group of solutions of $p^2 - cq^2 = 1$ that takes $*$-products to $*$-products).

8.3 The Quartic Pell's Equation with Negative Parameter

Exercise 3.1.

(a) Verify that $g_{-1}(x, y, z, w) = (x^2 - z^2 + 2yw)^2 + (y^2 - w^2 - 2xz)^2$, so that $g_{-1}(x, y, z, w) = 1$ requires either

$$x^2 - z^2 + 2yw = \pm 1, \quad y^2 - w^2 - 2xz = 0,$$

or

$$x^2 - z^2 + 2yw = 0, \quad y^2 - w^2 - 2xz = \pm 1.$$

Exercise 3.2. Let $c = -d$ for some positive integer $d \geq 2$.

(a) Verify that

$$g_c(x, y, z, w) = (x^2 - dz^2 + 2dyw)^2 + d(y^2 - dw^2 - 2xz)^2.$$

(b) Deduce that $g_c(x, y, z, w) = 1$ requires that

$$\begin{aligned} x^2 - dz^2 &= \pm 1 - 2dyw, \\ y^2 - dw^2 &= 2xz. \end{aligned}$$

Exercise 3.3. Verify that $g_{-2}(1, 4, 4, 2) = 1$ and use this fact to obtain other solutions of $g_{-2}(x, y, z, w) = 1$.

Exercise 3.4. Determine solutions of $g_{-3}(x, y, z, w) = 1$. Try either $x = 1$ or $w = 0$.

Exercise 3.5. When $x = 1$ and $x^2 - dz^2 + 2dyw = 1$, the equations in Exercise 4.2(b) become $z^2 = 2yw$ and $y^2 - dw^2 = 2z$. By trying values of (y, z, w), determine values of d for which a solution can be found.

Exercise 3.6. When $w = 0$, the equations in Exercise 3.2(b) become $x^2 - dz^2 = \pm 1$ and $y^2 = 2xz$. For what values of d can solutions to the system be found?

Exercise 3.7. When $y = 0$, the equations in Exercise 3.2(b) become $x^2 - dz^2 = \pm 1$ and $dw^2 = -2xz$. For what values of d can such solutions to the system be found?

8.4 The Quintic Pell's Equation

Exercise 4.1. Verify that

$$\begin{aligned} g_c(x, y, z, u, v) = {} & (x^5 + cy^5 + c^2z^5 + c^3u^5 + c^4v^5) \\ & - 5c(x^3yv + x^3zu + xy^3z) \\ & - 5c^2(y^3uv + xz^3v + yz^3u + xyu^3) \\ & - 5c^3(zu^3v + xuv^3 + yzv^3) + 5c(x^2y^2u + x^2yz^2) \\ & + 5c^2(x^2u^2v + x^2zv^2 + xy^2v^2 + xz^2u^2 + y^2z^2v + y^2zu^2) \\ & + 5c^3(yu^2v^2 + z^2uv^2) - 5c^2(xyzuv). \end{aligned}$$

Exercise 4.2. Prove that

$$\begin{aligned} g_{c^2}(x, y, z, u, v) &= g_c(x, cu, y, cv, z), \\ g_{c^3}(x, y, z, u, v) &= g_c(x, cz, c^2v, y, cu), \\ g_{c^4}(x, y, z, u, v) &= g_c(x, c^3v, c^2u, cz, y). \end{aligned}$$

Use these to determine solutions for $g_4(x, y, z, u, v) = 1$, $g_8(x, y, z, u, v) = 1$, and $g_{16}(x, y, z, u, v) = 1$ when it is given that $g_2(1, 1, 1, 1, 1) = g_2(1, 1, 0, 1, 0) = 1$.

Exercise 4.3. For $c = 2, 3, 4, 5$, try to determine some solutions of $g_c(x, y, z, u, v) = 1$ by inspection. Try solutions for which all the entries consist of $1, -1$, and 0. Use these solutions to obtain other solutions. Make use of a computer or pocket calculator.

8.5 The Sextic Pell's Equation

Exercise 5.1. Prove that in the sextic case

$$g_c(x, y, z, u, v, w) = p^2 - cq^2 = r^3 + cs^3 + c^2t^3 - 3crst,$$

where

$$\begin{aligned} p &= x^3 + (3xu^2 + 3y^2v + z^3 - 3xyw - 3xvz - 3uyz)c \\ &\quad + (v^3 + 3zw^2 - 3uvw)c^2, \\ q &= (3x^2u + y^3 - 3xyz) \\ &\quad + (u^3 + 3yv^2 + 3z^2w - 3xvw - 3uyw - 3uvz)c + w^3c^2, \\ r &= x^2 + 2czv - cu^2 - 2cyw, \\ s &= 2xz + cv^2 - y^2 - 2cuw, \end{aligned}$$

and

$$t = z^2 + 2xv - 2yu - cw^2.$$

Exercise 5.2.

(a) Suppose $x^2 - cu^2 = 1$. Determine $g_c(x, 0, 0, u, 0, 0)$.

(b) Suppose $x^3 + cz^3 + c^2v^3 - 3cxzv = 1$. Determine $g_c(x, 0, z, 0, v, 0)$.

(c) Use parts (a) and (b) to explicitly determine solutions for $g_c(x, y, z, u, v, w)=1$ for various values of c.

Exercise 5.3.

(a) Verify that

$$g_c(x, y, 0, 0, 0, 0) = x^6 - cy^6.$$

(b) Prove that $g_c(x, 2, 0, 0, 0, 0) = 1$ is solvable if and only if $x \equiv \pm 1 \pmod{32}$. What are the two lowest values of c that occur?

Exercise 5.4. Let $c = a^2$. Using the notation of Exercise 5.1, verify that

$$p + aq = (x + au)^3 + a(y + av)^3 + a^2(z + aw)^3 - 3a(x + au)(y + av)(z + aw).$$

8.6 Explorations

Exploration 8.1. The quartic equation $g_{15}(x, y, z, w) = 1$ is satisfied by $(x, y, z, w) = (31, 16, 8, 4)$, while $g_{17}(x, y, z, w) = 1$ is satisfied by $(x, y, z, w) = (33, 16, 8, 4)$. Try to generalize.

Exploration 8.2. Determine as many solutions in integers (x, y, z, w) as you can of $g_2(x, y, z, w) = 1$. How are they related with respect to $*$-multiplication? Are they all $*$-powers of the same solutions as in the quadratic and cubic cases? Look at other positive values of c and at rational solutions as well.

Exploration 8.3. The following solutions (x, y, z, w) of $g_2(x, y, z, w) = 1$ follow a pattern: (1, 2, 1, 0), (3, 2, 2, 2), (7, 6, 5, 4), (17, 14, 12, 10), (41, 34, 29, 24). Note also the solutions (x, y, z, w) of $g_3(x, y, z, w) = 1$: (2, 1, 1, 1), (7, 5, 4, 3), (26, 19, 15, 11), (97, 71, 56, 41). Investigate.

Exploration 8.4. Can one obtain solutions to $g_c(x, y, z, w) = 1$ by looking at suitable continued fraction expansions of θ where the convergent fractions are elements of $\mathbf{Z}(\sqrt{c})$? For example, it appears that the appropriate continued fractions for $2^{1/4}$ and $3^{1/4}$ are

$$2^{\frac{1}{4}} = 1 + 1/(2\sqrt{2} + 2) + 1/2 + 1/(2\sqrt{2} + 2) + 1/2 + \cdots$$

and

$$3^{\frac{1}{4}} = 1 + 1/(1 + \sqrt{3}) + 1/2 + 1/(1 + \sqrt{3}) + 1/2 + \cdots.$$

In a way analogous to the quadratic case, these generate Type A solutions.

Exploration 8.5. It is possible to get a Type B solution $(x, y, z, w) = (u, 0, v, 0)$ to $g_c(x, y, z, w) = 0$ from a solution to $u^2 - cv^2 = \pm 1$. Once a Type A solution is known, other Type B solutions for which y and w are not both equal to zero can be found by $*$—multiplication. Can we generate Type A solutions when Type B solutions are known? Given (x, y, z, w) for which $g_c(x, y, z, w) = 1$, show that $(x, -y, z, -w)$ also satisfies the same equation. How do the corresponding values of (p, q) relate? Look at $(x, y, z, w) * (x, -y, z, -w)^{-1}$.

Exploration 8.6. For various values of c, find nontrivial solutions to the equation $g_c(x, y, z, w) = 0$?

Exploration 8.7. Determine solutions of $g_c(x, y, z, w) = \pm 1$ for which c, x, y, z, w are polynomials in one or more variables.

Exploration 8.8. Investigate solutions of the sextic Pell's equation $g_2(x, y, z, u, v, w) = \pm 1$. As in the quartic case, these induce solutions of lower-degree Pell's equations, in this case

$$\xi^3 + c\eta^3 + c^2\zeta^3 - 3c\xi\eta\zeta = \pm 1$$

and $\rho^2 - c\sigma^2 = \pm 1$. What possible solutions of these two equations arise? Are there nontrivial solutions of the sextic that induce the trivial solutions $(1, 0, 0)$ and $(1, 0)$ of the cubic and quadratic, respectively?

Exploration 8.9. Investigate the solutions to the sextic Pell's equation when the parameter c is negative, or square, or cubic, or a sixth power.

Exploration 8.10. Let θ be an nth root of the integer c, and let r be a positive integer less than n. What is $N\big(1 + \theta + \theta^2 + \cdots + \theta^r\big)$?

Exploration 8.11. Study solutions of the nth-degree Pell's equation $g_2(x_1, x_2, \ldots, x_n) = 1$, and look for patterns.

Exploration 8.12. Let $n = dm$ for positive integers d and m. Let $\theta_n = c^{1/n}$ and $\theta_m = c^{1/m}$. Let N_n and N_m be their respective norms. Compare $N\big(x_1 + x_2\theta^d + \cdots + x_m\theta^{(m-1)d}\big)$ and $N\big(x_1 + x_2\theta_m + \cdots + x_m\theta_{m-1}^{m-1}\big)$. Use this to show how solutions of the nth-degree Pell's equation $g_c = 1$ can be found from solutions of the mth-degree Pell's equation.

8.7 Notes

In the determination of the units among the integers of the number fields, the reader will have noticed the situation becoming more complex with the degree of the Pell's equation. For the quadratic and cubic equations, the set of units is simply all the $*$-powers of a fundamental unit multiplied by ± 1. For the higher-degree cases, the structure is more elaborate. The situation is described by Dirichlet's Unit Theorem, a good account of which can be found in Chapter 2 of *Number Theory* by Borevich and Shafarevich; the theorem itself appears on page 112. About our present situation it says the following. Suppose we adjoin to the rationals **Q** the root θ of a polynomial equation with integer coefficients and leading coefficient equal to 1 to form the field $\mathbf{Q}(\theta)$. Suppose further that this polynomial has s real roots and $2t$ nonreal ones (i.e., t complex conjugate pairs of roots). Let $r = s + t - 1$. Then every unit among the integers of $\mathbf{Q}(\theta)$ can be written in the form

$$\zeta \epsilon_1^{a_1} \epsilon_2^{a_2} \cdots \epsilon_r^{a_r},$$

where ζ is a root of unity, the ϵ_i are fundamental units, and the a_i are integers. Thus, for the quadratic equation $x^2 - dy^2 = 1$ with d a positive nonsquare, $s = 2$ and $r = 1$. (When $d < 0$, we get $t = 1$ and $r = 0$ and there is only a finite set of units.) For the cubic equation $x^3 + cy^3 + c^2z^3 - 3cxyz = 0$ with c a noncube, $s = t = 1$ and $r = 1$. What can one expect for the quartic and higher-order Pell equations, and how can one find the fundamental units?

8.8 Hints

2.1. Write the factors in the form $(x + z\theta^2) + \theta(y + w\theta^2)$, etc., and arrange it so you are multiplying a sum by a difference to get a product that is a difference of squares.

2.5(b). Note that z and $1 + az$ constitute a coprime pair of integers whose product is a square.

2.18. Note that the following mapping preserves multiplication:

$$\begin{aligned} x + y\theta + z\theta^2 + w\theta^3 &= (x + z\theta^2) + \theta(y + w\theta^2) \\ &\longrightarrow [(x + z\theta^2) + \theta(y + w\theta^2)][(x + z\theta^2) = \theta(y + w\theta^2))] \\ &= (x + z\theta^2)^2 - \theta^2(y + w\theta^2)^2. \end{aligned}$$

and

$$p - q\sqrt{c} \rightarrow p + q\sqrt{\bar{c}}.$$

9

A Finite Version of Pell's Equation

The material in this chapter was developed by Jeffrey Higham in 1993, when he was an undergraduate at the University of Toronto. We have already seen that with respect to the multiplication

$$(x, y) * (u, v) = (xu + dyv, xv + yu)$$

defined for pairs of integers, the solutions of Pell's equation $x^2 - dy^2 = 1$ can be obtained by taking "powers" of an elementary solution. We will explore this structure when we consider instead pairs (x, y) for which $x^2 - dy^2$ differs from 1 by a multiple of some fixed modulus m.

This chapter will begin with a particular case to indicate the setting, and then look at the situation where the modulus is a prime or prime power.

9.1 Solutions Modulo 11

We recall the notion of modular arithmetic introduced in Section 4.1, and take the modulus to be 11. Imagine a clock for a day that has only 11 hours. Then two hours after ten o'clock, it will be one o'clock; five hours after nine o'clock, it will be three o'clock. Working modulo 11, we consider only the numbers $0, 1, 2, \ldots, 10$. The sum of two of these numbers with our new arithmetic is the ordinary sum when that sum does not exceed 10 and the ordinary sum minus 11 otherwise; thus the sum is also a member of the set.

We can multiply two numbers, modulo 11, in a similar way. The product is the ordinary product minus the highest multiple of 11 that does not exceed it. Since, for example, $5 \times 6 = 30 = 22 + 8$, we write

$$5 \times 6 \equiv 8 \pmod{11}$$

(read, "5×6 is congruent to 8 modulo 11") to express this new type of product. For sums, we can write, for example,

$$3 + 9 \equiv 1 \pmod{11}$$

(read, "3+9 is congruent to 1 modulo 11").

Exercises 1.1. Verify that $3+8 \equiv 0 \pmod{11}$, $5-9 \equiv 7 \pmod{11}$ and $6 \times 8 \equiv 4 \pmod{11}$. Construct an addition table and a multiplication table, modulo 11; the rows and columns will be headed by the numbers 0, 1, 2, ..., 9, 10.

Exercise 1.2. Verify that $5^2 - 2 \times 4^2 \equiv 4$ and $4^2 - 3 \times 4^2 \equiv 1 \pmod{11}$.

Exercise 1.3. Make a table with columns headed by x, x^2, $2x^2$, $3x^2$, and $4x^2$, where x assumes the values 0, 1, 2, ..., 9, 10. Fill in the values of these functions of x, where the square is evaluated modulo 11. Use this table to find solutions to the following congruences:

$$\begin{aligned} x^2 - y^2 &\equiv 1, \\ x^2 - 2y^2 &\equiv 1, \\ x^2 - 3y^2 &\equiv 1, \\ x^2 - 4y^2 &\equiv 1, \end{aligned}$$

modulo 11.

Exercise 1.4. Let d be a positive integer. As was done with the regular Pell's equation, we can define a product of pairs by

$$(x, y) * (u, v) \equiv (z, w),$$

where $z \equiv xu + dyv$ and $w \equiv xv + yu \pmod{11}$. Verify that when $d = 2$,

$$(3, 7) * (2, 9) \equiv (0, 8).$$

Exercise 1.5. In the notation of Exercise 1.4, verify that if $x^2 - dy^2 \equiv 1$ and $u^2 - dv^2 \equiv 1$, then $z^2 - dw^2 \equiv 1$, modulo 11.

Exercise 1.6. Determine all the solutions of the congruence

$$x^2 - 2y^2 \equiv 1 \pmod{11}$$

with $0 \le x, y \le 10$. How many solutions are there? Construct a multiplication table for the solutions such that the entry in the row headed by (r, s) and the column headed by (u, v) is $(r, s) * (u, v)$.

Exercise 1.7. Define $(u, v)^m$ to be the $*$-product of m terms, each equal to (u, v). Find a solution (u, v) of $x^2 - 2y^2 \equiv 1 \pmod{11}$ such that every solution can be written in the form $(u, v)^m$ for some nonnegative integer m.

Exercise 1.8. Do Exercises 1.6 and 1.7 with the number 2 replaced by 1, 3, and 4. What property do 1, 3, and 4 share that 2 fails to have?

9.2 The Case of a Prime Modulus

Let p be an odd prime number. Denote by $\mathbf{Z}_p$ the set $\{0, 1, 2, \ldots, p-1\}$. Addition and multiplication are defined in $\mathbf{Z}_p$ from ordinary addition and multiplication, modulo p; in other words, replace the ordinary sum and product by their remainders upon division by p. In $\mathbf{Z}_p$ with these operations, we have access to the usual arithmetic laws of commutativity, associativity, and distributativity. But furthermore, we have a kind of division available to us, since given any nonzero element a in $\mathbf{Z}_p$, it turns out that there is an element u for which $au \equiv 1$; this *reciprocal* u will be denoted by a^{-1}. Given a fixed positive integer d, we can define a *-product of pairs by the formula

$$(x, y) * (u, v) \equiv (xu + dyv, xv + yu),$$

and it can be checked that if $(x, y) = (r, s)$, (u, v) both satisfy $x^2 - dy^2 \equiv 1 \pmod{p}$, then so also does $(x, y) \equiv (r, s) * (u, v)$. Note that $(1, 0) * (u, v) = (u, v)$ for each pair (u, v), so that $(1, 0)$ is the identity element for the *-product. Each element (u, v) for which $u^2 - dv^2 \equiv 1 \pmod{p}$ also has an "inverse" for this product, namely $(u, p - v)$ (i.e., $(u, v) * (u, p - v) \equiv (1, 0)$).

Let $G(p, d)$ denote the set of distinct solutions, modulo p, of the congruence $x^2 - dy^2 \equiv 1 \pmod{p}$. In the first section we found that for $1 \le d \le 4$, $G(11, d)$ has either 10 or 12 members, and that the solutions were all *-powers of a particular solution. One might also observe that 1, 3, and 4 are all squares modulo 11, while 2 is not. In this section we will explore the significance of this observation in general.

Exercises 2.1. Suppose that $d \equiv r^2 \pmod{p}$ for some integer r. Verify that $x^2 - dy^2 \equiv (x - ry)(x + ry) \pmod{p}$. Thus, solving $x^2 - dy^2 \equiv 1$ involves determining pairs of reciprocals modulo p. [Note that u and v are *reciprocals* modulo p if and only if u and v are not divisible by p and $uv \equiv 1 \pmod{p}$.]

Exercise 2.2. To find the reciprocal of a number a, modulo p, we must solve a congruence of the form

$$ax \equiv b \pmod{p}, \tag{1}$$

where in finding the reciprocal, $b = 1$. We will examine the solvability of such congruences.

(a) Solve the following congruences modulo 11: $5x \equiv 8$, $3x \equiv 1$, $7x \equiv 0$, $0x \equiv 5$, $0x \equiv 0$.

(b) For a general prime modulus p, discuss the solvability of $0x \equiv b$ in the cases where $b \equiv 0$ and $b \not\equiv 0$.

(c) Suppose that $a \not\equiv 0 \pmod{p}$. Prove that $ax \equiv 0 \bmod p$ has only the solution $x \equiv 0$, and deduce that as x runs through the integers $0, 1, 2, \ldots, p-1$, then ax runs through the same integers in some order, each integer occurring exactly once. From this, conclude that $ax \equiv b$ has exactly one solution, modulo p, for each integer b, and that in particular, a has a unique reciprocal modulo p.

Exercise 2.3. Return to the congruence $x^2 - r^2y^2 \equiv 1$, where $r \not\equiv 0 \pmod{p}$. Suppose u and v are reciprocals modulo p, i.e., u and v are not congruent to 0 and $uv \equiv 1$. Prove that the system of two congruences

$$x - ry \equiv u,$$
$$x + ry \equiv v,$$

is uniquely solvable for x and $y \pmod{p}$, and deduce that there is a one-to-one correspondence between solutions of $x^2 - r^2y^2 \equiv 1$ and pairs (u, v) of reciprocals, modulo p. Conclude that $\#G(p, r^2) = p - 1$. ($\#S$ means "the number of elements in the set S".)

Exercise 2.4. There is a one-to-one correspondence between solutions (x, y) of Pell's congruence $x^2 - r^2y^2 \equiv 1 \pmod{p}$ and nonzero integers w given by

$$(x, y) \sim w \equiv x - ry.$$

(a) Prove that if $(x_1, y_1) \sim w_1$ and $(x_2, y_2) \sim w_2$, then $(x_1, y_1) * (x_2, y_2) \sim w_1w_2$ where the product w_1w_2 is taken in $\mathbf{Z}_p$.
(b) Prove that every solution of $x^2 - r^2y^2 \equiv 1 \pmod{p}$ is a *-power of some fundamental solution (x_1, y_1) if and only if $(x_1 - ry_1)^k$ runs through all the nonzero elements of $\mathbf{Z}_p$ as k runs through the positive integers. (Such a number $x_1 - ry_1$ is called a *primitive root* modulo p.) ♠

When d is not congruent to a square, modulo p, we cannot make use of a factorization of $x^2 - dy^2$ to study the solutions of the Pell's congruence $x^2 - dy^2 \equiv 1$. However, this strategy can be adapted. For any nonzero element a in $\mathbf{Z}_p$, we define a^{-1} to be the reciprocal of a, i.e., the unique element c for which $ac \equiv 1 \pmod{p}$. Such an element is called the (multiplicative) *inverse* a.

Exercise 2.5.
(a) Prove that $(ab)^{-1} \equiv b^{-1}a^{-1} \pmod{p}$.
(b) Suppose that $x^2 - dy^2 \equiv 1 \pmod{p}$ with $y \not\equiv 0$. Verify that this is equivalent to

$$(xy^{-1} - y^{-1})(xy^{-1} + y^{-1}) \equiv d \pmod{p}.$$

(c) Suppose that $xy^{-1} - y^{-1} \equiv w$. Verify that w is a nonzero element in $\mathbf{Z}_p$ and that $xy^{-1} + y^{-1} \equiv dw^{-1}$.
(d) Given a nonzero element w in $\mathbf{Z}_p$, we wish to show that the system

$$\begin{aligned} xy^{-1} - y^{-1} &\equiv w, \\ xy^{-1} + y^{-1} &\equiv dw^{-1}, \end{aligned} \tag{1}$$

is uniquely solvable for x and y. Rewrite the system as

$$x - wy \equiv 1,$$
$$x - dw^{-1}y \equiv -1.$$

This is a linear system of congruences in x and y that can be solved by eliminating one of the variables. Verify that elimination of y leads to

$$(d - w^2)x \equiv d + w^2,$$

while elimination of x leads to

$$(d - w^2)y \equiv 2w.$$

(e) Suppose that d is not the square of any element in $\mathbf{Z}_p$. Argue that the system (1) is solvable for any nonzero element w and that such a solution is given by

$$x \equiv (d + w^2)(d - w^2)^{-1},$$
$$y \equiv 2w(d - w^2)^{-1}.$$

Deduce that there are $p - 1$ solutions (x, y) of Pell's congruence for which $y \not\equiv 0$.

(f) Suppose that $d \equiv r^2$ for some nonzero element r in $\mathbf{Z}_p$. Let $s \in \mathbf{Z}_p$ be such that $s^2 \equiv r^2$. Show that $0 \equiv (s - r)(s + r)$ and deduce that p divides either $s - r$ or $s + r$, so that the only square roots of d modulo p are r and $p - r$.

(g) Prove that when $d \equiv r^2$, the system (1) is solvable for x and y for any nonzero w in $\mathbf{Z}_p$ except for r and $p - r$. Deduce that there are $p - 3$ solutions (x, y) for Pell's congruence for which $y \not\equiv 0$.

Exercise 2.6. Go through the process of Exercise 2.5 to find solutions of the congruences $x^2 - 2y^2 \equiv 1$ and $x^2 - 3y^2 \equiv 1$, modulo 11.

Exercise 2.7. The only remaining case to consider is $y \equiv 0$, and then we can count the solutions to Pell's equation.

(a) Show that if $x^2 - dy^2 \equiv 1$ and $y \equiv 0 \pmod{p}$, then $x \equiv 1$ or $x \equiv p - 1$.

(b) Deduce from (a) and Exercises 2.3 and 2.5 that the number of elements of $G(p, d)$ is given by

$$\#G(p, d) = \begin{cases} p - 1, & \text{if } d \text{ is a nonzero square (mod } p\text{);} \\ p + 1, & \text{if } d \text{ is not a square (mod } p\text{);} \\ 2, & \text{if } d \equiv 0 \pmod{p}. \end{cases}$$

9.3 The Structure of Z_p

We digress from our study of Pell's congruence to record some basic facts about $\mathbf{Z}_p$.

A *polynomial over* $\mathbf{Z}_p$ is an expression of the form

$$a_n t^n + a_{n-1} t^{n-1} + \cdots + a_1 t + a_0,$$

where t is a variable and a_i belongs to $\mathbf{Z}_p$ for each i. When $a_n \not\equiv 0$, n is the *degree* of the polynomial.

Our goal is to show that there is a nonzero element of $\mathbf{Z}_p$ such that every nonzero element of this set is a power of it.

Exercises 3.1. Suppose that a, b and c are any elements of $\mathbf{Z}_p$.

(a) Verify that $a + b \equiv b + a$, $ab \equiv ba$, $(a + b)c \equiv ac + bc$, $(a + b) + c \equiv a + (b + c)$, and $(ab)c \equiv a(bc)$.

(b) Write $-a$ for $p - a$. Verify that $a + (-a) \equiv 0$.

(c) Verify that $0 + a \equiv a$, $0 \cdot a \equiv 0$, and $1 \cdot a \equiv a$.

(d) Suppose that $ab \equiv 0$. Prove that either $a \equiv 0$ or $b \equiv 0$.

Exercise 3.2. Let $f(t)$ be a polynomial over $\mathbf{Z}_p$, and suppose that $a \in \mathbf{Z}_p$. In exactly the same way as for polynomials with real coefficients, we can divide $f(t)$ by $t - a$ to get a presentation of the form

$$f(t) \equiv (t - a)q(t) + r,$$

where r is an element of $\mathbf{Z}_p$ called the *remainder*.

(a) Verify that in $\mathbf{Z}_{11}$,

$$t^3 + 7t^2 + 2t + 3 \equiv (t - 5)(t^2 + t + 7) + 5.$$

(b) *Remainder theorem.* Prove that the remainder when $f(t)$ is divided by $t - a$, is congruent to $f(a)$.

(c) *Factor theorem.* Prove that a is a *root* of $f(t)$ modulo p (i.e., $p(a) \equiv 0$) if and only if $t - a$ is a factor of $f(t)$.

(d) Suppose that k is the degree of the polynomial $f(t)$ and that $a_1, a_2, \ldots, a_r$ are pairwise incongruent roots of $f(t)$. Prove that

$$f(t) \equiv (t - a_1)(t - a_2) \cdots (t - a_r)q(t)$$

for some polynomial $q(t)$ of degree $k - r$, so that in particular, the number of incongruent roots of $f(t)$ cannot exceed k.

Exercise 3.3. *Little Fermat Theorem.* Observe that when $a \not\equiv 0 \pmod{p}$, the set $\{a, 2a, 3a, \ldots, (p - 1)a\}$ is, modulo p, a rearrangement of the set $\{1, 2, 3, \ldots, p - 1\}$. Deduce that

$$(p - 1)!a^{p-1} \equiv (p - 1)! \pmod{p}$$

and that $a^{p-1} \equiv 1 \pmod{p}$.

Exercise 3.4. Prove that $t^{p-1} - 1 \equiv (t - 1)(t - 2) \cdots (t - \overline{p - 1})$.

Exercise 3.5.

(a) For each of the nonzero members a in $\mathbf{Z}_{11}$ determine the smallest number k for which $a^k \equiv 1 \pmod{11}$.

(b) Answer (a) with 11 replaced by 13.

(c) In (a) and (b), how is k related to the primes 11 and 13, respectively?

Exercise 3.6. Let $a \not\equiv 0 \pmod{p}$, and suppose that k is the smallest positive integer for which $a^k \equiv 1 \pmod{p}$. We say that "*a belongs to the exponent k modulo p*." Prove that k is a divisor of $p - 1$. ♠

In the next few exercises we show that whenever k is a positive divisor of $p - 1$, there is some number $a \in \mathbf{Z}_p$ for which $a^k \equiv 1$. To attain this, we define *Euler's totient function* $\phi(m)$.

For any positive integer m, $\phi(m)$ is the number of positive integers x not exceeding m for which the greatest common divisor of x and m is equal to 1 (i.e., x and m are *relatively prime*).

Exercise 3.7.
(a) Verify that $\phi(2^r) = 2^{r-1}$ for positive integer r.
(b) Verify that for an odd prime p, $\phi(p) = \phi(2p) = p - 1$ and $\phi(p^r) = p^r - p^{r-1}$ when $r \geq 2$.
(c) Verify that $\phi(6) = 2$ and that $\phi(10) = \phi(12) = 4$.
(d) When does $\phi(m)$ assume an odd value?

Exercise 3.8. Verify that

$$12 = \phi(1) + \phi(2) + \phi(3) + \phi(4) + \phi(6) + \phi(12).$$

Make a conjecture and test it for integers other than 12.

Exercise 3.9. Prove that for each positive integer m,

$$m = \sum\{\phi(k) : k \text{ is a positive divisor of } m\}.$$

Exercise 3.10. Let k be a positive divisor of $p - 1$.
(a) Prove that $t^{p-1} - 1 \equiv (t^k - 1)q(t)$, where $q(t)$ is a polynomial over $\mathbf{Z}_p$ of degree $p - 1 - k$ with $(p - 1)/k$ terms, each a power of t^k.
(b) Observe that $t^{p-1} - 1$ has exactly $p - 1$ roots in $\mathbf{Z}_p$, while $q(t)$ has at most $p - 1 - k$ roots. If $a^k \equiv 1$, show that $q(a) \not\equiv 0$. Deduce that there are exactly k elements of $\mathbf{Z}_p$ for which $a^k \equiv 1$.

Exercise 3.11. Suppose that $a \in \mathbf{Z}_p$ and a belongs to the exponent k. Prove that there are exactly $\phi(k)$ powers of a that belong to the exponent k, and that these are the only elements of $\mathbf{Z}_p$ that belong to the exponent k.

Exercise 3.12. Suppose that k is a divisor of $p - 1$. Define $\psi(k)$ to be the number of elements in $\mathbf{Z}_p$ that belong to the exponent k. Observe, that by Exercise 3.11, if $\psi(k) > 0$, then $\psi(k) = \phi(k)$.
(a) Prove that

$$p - 1 = \sum\{\psi(k) : k \text{ is a positive divisor of } p\}.$$

(b) Note that $\psi(k) \leq \phi(k)$ for each k, and deduce from Exercise 3.9 and (a) that in fact $\psi(k) = \phi(k)$ for every divisor k of $p - 1$.

(c) In particular, deduce that $\mathbf{Z}_p$ contains exactly $\phi(p-1)$ elements g that belong to the exponent $p-1$. Such an element g is called a *primitive root* modulo p.

Exercise 3.13.
(a) Prove that $\mathbf{Z}_p = \{0, g, g^2, \ldots, g^{p-1}\}$ whenever g is a primitive root.
(b) Determine all the primitive roots modulo 3, 5, 7, 11, and 13.

Exercise 3.14. Suppose that r is a nonzero element of $\mathbf{Z}_p$. Prove that there is a solution $(x, y) \equiv (u, v)$ of Pell's congruence $x^2 - r^2y^2 \equiv 1$ such that all the distinct solutions are given by

$$(x, y) \equiv (u, v)^k,$$

where $0 \leq k \leq p - 1$.

Exercise 3.15. We can use primitive roots to provide an explicit representation of the solutions of $x^2 - r^2y^2 \equiv 1$. Select $(x_1, y_1) \in G(p, r^2)$ such that

$$x_1 + ry_1 \equiv g,$$
$$x_1 - ry_1 \equiv g^{-1}.$$

(a) Verify that $x_1 = 2^{-1}(g + g^{-1})$ and $y_1 = (2r)^{-1}(g - g^{-1})$.
(b) If $(x_n, y_n) = (x_1, y_1)^n$, prove that

$$x_n \equiv 2^{-1}(g^n + g^{-n}),$$
$$y_n \equiv (2r)^{-1}(g^n - g^{-n}),$$

where g^{-n} means $(g^{-1})^n$, and is the inverse of g^n.
(c) Check (a) and (b) in the special cases that p is equal to 11 and 13.

9.4 The Structure of $G(p, d)$

As in the previous section, p is an odd prime and d is a positive integer not exceeding $p - 1$. This section relies on results about Chebyshev polynomials obtained in Section 3.4.

Exercises 4.1. Suppose that $x^2 - dy^2 \equiv 1$ is satisfied by $(x, y) \equiv (u, v)$. Define $(u_n, v_n) \equiv (u, v)^n$ for each positive integer n. Prove that

$$(x_n, v_n) \equiv (T_n(u), vU_n(u)),$$

where T_n and U_n denote Chebyshev polynomials of the first and second kinds. Thus, given a solution of Pell's congruence, we have obtained a representation of other solutions.

Exercise 4.2. Verify that the solutions of $x^2 - 2y^2 \equiv 1 \bmod 11$ are given by $(x, y) \equiv (T_n(3), 2U_n(3))$, where n runs over the positive integers from 1 to 12, inclusive.

Exercise 4.3. The set $G(p, d)$ is a finite set of pairs of elements from $\mathbf{Z}_p$. Let (u, v) be one of these.

(a) With $-v$ denoting $p - v$, verify that $(u, -v) \in G(p, d)$ and that $(u, v) * (u, -v) \equiv (1, 0)$.

(b) Observe that at most finitely many powers $(u, v)^k$ are distinct, and deduce that there is a positive integer m for which $(u, v)^m \equiv (1, 0)$.

(c) Let (u, v) belong to the exponent m, so that m is the smallest positive integer for which $(u, v)^m \equiv (1, 0)$. Define $(u_i, v_i) \equiv (u, v)^i$ for $1 \le i \le m$. Prove that each u_i occurs exactly twice unless $u_i \equiv 1$ or $u_i \equiv p - 1$, and that $v_{m-i} \equiv -v_i$.

(d) With the notation of (c), verify that $(u_i, v_i)^m \equiv (1, 0)$. ♠

The result that we are heading for is that $G(p, d)$ consists of the *-powers of a "generating element." We shall follow the strategy used in Section 9.3 to show that there is a primitive root modulo p. We show that the number of elements of $G(p, d)$ belonging to the exponent m does not exceed $\phi(m)$ for each m.

Exercise 4.4. Suppose that $(u, v) \in G(p, d)$ has $v \not\equiv 0$ and belongs to an even exponent $2k$.

(a) Deduce that $T_{2k}(u) \equiv 1$ and so, from Exercise 3.4.6, $U_k(u) \equiv 0$.

(b) Regard $U_k(t)$ as a polynomial over $\mathbf{Z}_p$. Observe that its degree is exactly $k - 1$ and deduce that it has at most $k - 1$ roots in $\mathbf{Z}_p$.

(c) With $(u_i, v_i) \equiv (u, v)^i$ for $1 \le i \le 2k$, prove that $T_{2k}(u_i) \equiv 1$ for each i, so that either $u_i \equiv 1$, $u_i \equiv p - 1$, or u_i is a root of $U_k(t)$. Deduce that each root of $U_k(t)$ appears among the u_i.

(d) Show that the only elements in $G(p, d)$ that belong to the exponent $2k$ are (u_i, v_i), where $\gcd(i, 2k) = 1$.

Exercise 4.5. Suppose the $(u, v) \in G(p, d)$ has $v \not\equiv 0$ and belongs to an odd exponent $2k + 1$.

(a) Deduce that $(U_{k+1} + U_k)(u) \equiv 0$.

(b) Regarding $(U_{k+1} + U_k)(t)$ as a polynomial over $\mathbf{Z}_p$, prove that its degree is exactly k and that it has at most k roots in $\mathbf{Z}_p$.

(c) With $(u_i, v_i) \equiv (u, v)^i$ for $1 \le i \le 2k$, prove that $T_k(u_i) \equiv 1$ for each i, so that $(U_{k+1} + U_k)(u_i) \equiv 0$ for each i.

(d) Deduce that each root of $(U_{k+1} + U_k)(t)$ appears among the u_i.

(e) Show that the only elements in $G(p, d)$ that belong to the exponent $2k + 1$ are (u_i, v_i), where $\gcd(i, 2k + 1) = 1$.

Exercise 4.6. Deduce from Exercises 4.4 and 4.5 that if for some positive integer m there is a solution $(u, v) \in G(p, d)$ that belongs to the exponent m, then there are exactly $\phi(m)$ such solutions, namely $(u, v)^i$ with i and m relatively prime.

Exercise 4.7. We show that if $(u, v) \in G(p, d)$ belongs to the exponent m, then m must divide $\#G(p, d)$, the number of elements in $G(p, d)$. Let S_1 be the set

$$\{(u, v)^i : 1 \leq i \leq m\}.$$

We form a finite family of sets $S_1, S_2, \ldots, S_h$ that partition $G(p, d)$ as follows.

Suppose that $S_1, S_2, \ldots, S_i$ have already been selected. If they do not exhaust $G(p, d)$, select an element (a, b) of $G(p, d)$ not already included in any of the sets chosen so far and define

$$S_{i+1} = \{(a, b) * (u, v)^j : 1 \leq j \leq m\}.$$

Observe that (a, b) belongs to S_{i+1}.

(a) Go through this process in the case of $G(11, 2)$ and the elements $(u, v) = (0, 4)$ and $(u, v) = (5, 1)$.
(b) Prove that each S_i contains m distinct objects.
(c) Prove that the S_i are pairwise disjoint.
(d) Note that $G(p, d)$ is the union of finitely many pairwise disjoint sets S_i, each with the same number of elements, and deduce that m divides $G(p, d)$.

Exercise 4.8. Suppose that $G(p, d)$ has n elements. For each positive integer m, let $f(m)$ be the number of elements of $G(p, d)$ that belong to the exponent m, so that in particular, $f(m) = 0$ when m is not a divisor of $p - 1$.
(a) Prove that $f(1) = f(2) = 1$.
(b) Prove that

$$n = \sum \{f(m) : m \quad \text{is a divisor of} \quad n\}.$$

(c) Deduce from (b) and Exercise 4.6 that $f(m) = \phi(m)$ for each divisor of n.
(d) Prove that $G(p, d) = \{(u, v)^i : 1 \leq i \leq n\}$ for a suitable one (u, v) of its elements.

9.5 Pell's Congruence Modulo a Prime Power

Let p be a prime and a any positive integer. We now turn to the set $G(p^a, d)$ of incongruent solutions to Pell's congruence $x^2 - dy^2 \equiv 1$ modulo the power p^a of a prime p.

Exercise 5.1.
(a) Determine the set $G(3, 2)$.
(b) Consider the congruence $x^2 - 5y^2 \equiv 1 \pmod 9$. Verify that each solution in fact satisfies $x^2 - 2y^2 \equiv 1 \pmod 3$.
(c) Write down, modulo 9, all the solutions of $x^2 - 2y^2 \equiv 1 \pmod 3$ and check which of them satisfy $x^2 - 5y^2 \equiv 1 \pmod 9$.
(d) Determine $G(9, 2)$.
(e) Determine $G(27, 2)$ and $G(27, 5)$.

Exercise 5.2. Let p be an odd prime and let $a \geq 2$. Any solution of $x^2 - dy^2 \equiv 1 \pmod{p^a}$ must satisfy $x^2 - dy^2 \equiv 1 \pmod{p^{a-1}}$. Accordingly, to find $G(p^a, d)$, we should first determine $G(p^{a-1}, d)$ and then see which solutions "can be lifted" to solutions in $G(p^a, d)$. Suppose that

$$u^2 - dv^2 \equiv 1 \pmod{p^{a-1}}.$$

This means that there is an integer c such that $u^2 - dv^2 = 1 + cp^{a-1}$. Let $w = u + sp^{a-1}$ and $z = v + tp^{a-1}$ for integers s and t. Then $w^2 - dz^2 \equiv 1 \pmod{p^{a-1}}$.

(a) Determine conditions on s and t to ensure that

$$w^2 - dz^2 \equiv 1 \pmod{p^a}.$$

(b) Prove that there are p incongruent pairs (s, t), modulo p, such that the condition in (b) is satisfied.

(c) Deduce that $\#G(p^a, d) = p\#G(p^{a-1}, d) = p^{a-1}\#G(p, d)$.

9.6 Explorations

Exploration 9.1. What can be said about the number of elements in $G(2^a, d)$? Does $G(2^a, d)$ even contain an element whose $*$-powers constitute the whole of the set?

Exploration 9.2. Examine the structure of $G(p^a, d)$ for p an odd prime and $a \geq 2$. For the case $a = 2$ and d a perfect square, it can be shown that $G(p^2, d)$ consists of all the *-powers of one of its elements by reference to the primitive root theorem for squares of primes. The primitive root theorem says that there is a number g such that each number m relatively prime to p is congruent to some power g^k of $g \pmod{p^2}$. There are $\phi(p^2) = p^2 - p$ numbers m relatively prime to p for which $1 \leq m \leq p^2 - 1$, and these can be put into one-to-one correspondence with the $p^2 - p$ elements of $G(p^2, d)$, following the strategy of Exercises 2.4 and 3.14.

A tool for accessing higher powers of p is Hensel's lemma. Let p be a prime and let $f(t)$ be a polynomial with integer coefficients. Suppose u_1 is an integer for which $f(u_1) \equiv 0$ and $f'(u_1) \not\equiv 0$ modulo p. Then for each positive integer k, there is a unique integer u_k satisfying $0 \leq u_k < p^k$, $u_k \equiv u_1 \pmod{p}$, and $f(u_k) \equiv 0 \pmod{p^k}$.

The theory of primitive roots, modulo a prime, can be found in many standard number theory texts. For a brief introduction to Hensel's lemma, consult E.J. Barbeau, *Polynomials*, Springer, New York, 1989, 1995, Exploration E.33 in Section 3.4.

Exploration 9.3. The fundamental solution of $x^2 - 2y^2 = 1$ is $(3, 2)$, and the complete set of solutions of the congruence with x positive is given by the $*$-powers of $(3, 2)$. If we reduce these solutions, modulo m, for which values of m do we obtain a complete set of solutions of the congruence $x^2 - 2y^2 \equiv 1 \pmod{m}$. If m is prime, to which exponent does (3, 2) belong? Look at this issue when 2 is replaced by other nonsquare numbers d.

9.7 Hints

2.3. Add and subtract the two congruences to get congruences that each involve just one variable. Proceed as though you were solving ordinary equations.

3.4. By Exercise 3.3, each nonzero element of $\mathbf{Z}_p$ is a root of $t^{p-1} - 1$. Now use the factor theorem.

3.6. Write $p - 1 = kq + r$ where, q and r are nonnegative integers and $0 \leq r \leq k - 1$. Use the fact that k is minimal and $a^{p-1} \equiv (a^k)^q a^r$.

3.9. The idea is to regard the right side of the equation as a count of the set of integers $\{1, 2, \ldots, m - 1, m\}$. Let u be one of these and let k be the greatest common divisor of u and m. Observe that u/k and m/k are relatively prime. Note that $\sum \phi(k) = \sum \phi(m/k)$ with both sums taken over the positive divisors of m.

3.11. k must be a divisor of $p - 1$, so $t^k - 1$ has exactly k roots. Prove that these must be $1, a, a^2, \ldots, a^{k-1}$ (check that these are all incongruent). Suppose $b = a^r$ is one of these. Prove that k is the smallest exponent for which $b^k \equiv 1$ if and only if r and k are relatively prime.

5.3(b). Argue that there must be two equal *-powers $(u, v)^r$ and $(u, v)^s$ with $0 < r < s$. Multiply each side of the equation $(u, v)^r = (u, v)^s$ r times by $(u, -v)$.

Answers and Solutions

Chapter 1

1.1(a). The point E can be obtained by folding the square along the bisector of angle BAC, so that AB is folded onto the diagonal with B landing on E. If F is the point where the bisector intersects BC, then BF folds to EF. Thus $EF \perp AC$ and $BF = EF$. Since $\angle EFC = 45° = \angle ECF$, $FE = EC$.

1.1(b). $|FC| = |BC| - |BF| = |AE| - |CE| = 2|AE| - |AC|$ and $|EC| = |AC| - |AE| = |AC| - |BC|$.

1.3(a).

n	p_n	q_n	r_n
1	1	1	1
2	3	2	$3/2 = 1.5$
3	7	5	$7/5 = 1.4$
4	17	12	$17/12 = 1.416666$
5	41	29	$41/29 = 1.413793$

1.3(c). Clearly, $r_n > 1$ for each n, so that from (b), $r_{n+1} - r_n$ and $r_n - r_{n-1}$ have opposite signs and

$$|r_{n+1} - r_n| < \frac{1}{4}|r_n - r_{n-1}|.$$

Since $r_2 > 1 = r_1$, it follows that $r_1 < r_3 < r_2$. Suppose as an induction hypothesis that

$$r_1 < r_3 < \cdots < r_{2m-1} < r_{2m} < \cdots < r_2.$$

Then $0 < r_{2m} - r_{2m+1} < r_{2m} - r_{2m-1}$ and $0 < r_{2m+2} - r_{2m+1} < r_{2m} - r_{2m+1}$, so that $r_{2m-1} < r_{2m+1} < r_{2m+2} < r_{2m}$. Let k, l be any positive integers. Then, for $m > k, l$, $r_{2k+1} < r_{2m-1} < r_{2m} < r_{2l}$.

1.3(d). Let α be the least upper bound (i.e., the smallest number at least as great as all) of the numbers in the set $\{r_1, r_3, r_5, \ldots, r_{2k+1}, \ldots\}$. Then $r_{2m-1} \leq \alpha \leq r_{2m}$

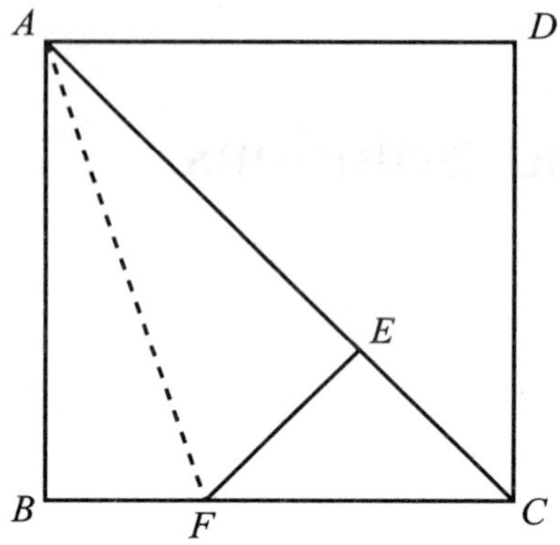

FIGURE 1.3.

for each positive integer m, so that $|\alpha - r_{2m-1}|$ and $|\alpha - r_{2m}|$ are both less than

$$|r_{2m} - r_{2m-1}| \leq \frac{1}{4^{2(m-1)}} |r_2 - r_1| = \frac{1}{2^{4m-3}}.$$

It follows that $\lim_{n\to\infty} r_n = \lim_{n\to\infty} r_{n-1} = \alpha$. Taking limits in the recursion in (d) leads to

$$\alpha = 1 + \frac{1}{1+\alpha},$$

which reduces to $\alpha^2 = 2$. Since $r_n > 0$ for each n, $\alpha > 0$.

1.4(c). $347/19 = 18 + 1/3 + 1/1 + 1/4$.

1.4(d). Since $\sqrt{2} = 1 + (\sqrt{2} - 1) = 1 + 1/1 + \sqrt{2}$, it follows that the process does not terminate, since we can always replace the final $1 + \sqrt{2}$ by $2 + 1/1 + \sqrt{2}$ to get the continued fraction in Exercise 1.3.

1.7(b). This can be established from Exercise 1.5(a) by induction.

1.7(c). For each positive integer k, let $p_{2k} = x_k$ and $q_{2k} = 2y_k$. Then, from Exercise 1.5(b), $x_k^2 - 8y_k^2 = p_{2k}^2 - 2q_{2k}^2 = 1$.

1.7(d). If $x^2 - 8y^2 = -1$, then x would have to be odd. But then $x^2 \equiv 1 \pmod 8$, yielding a contradiction.

2.4(a). Since $q_n + q_{n-1} = (q_n - q_{n-1}) + 2q_{n-1} = p_{n-1} + (p_n - p_{n-1}) = p_n$, $p_n(q_n - q_{n-1}) = p_n p_{n-1} = p_{n-1}(q_n + q_{n-1})$.

2.4(b). From Exercise 2.3(c),

$$\begin{aligned} p_{n+1}q_{n+1} &= 4p_nq_n + 2(p_nq_{n-1} + p_{n-1}q_n) + p_{n-1}q_{n-1} \\ &= 4p_nq_n + 2(p_nq_n - p_{n-1}q_{n-1}) + p_{n-1}q_{n-1} = 6p_nq_n - p_{n-1}q_{n-1}. \end{aligned}$$

3.1(b). Yes. It suffices to consider the case for which the greatest common divisor of a, b, c is 1. Since each odd square leaves a remainer 1 and no square leaves a remainder 2 upon division by 4, both of a and b cannot be odd. Suppose b is even and a is odd. Then $b^2 = (c-a)(c+a)$ expresses b as the product of two even divisors whose greatest common divisor is 2. Thus, there are integers m and n for which $c+a = 2m^2$ and $c-a = 2n^2$, and the result holds.

3.2(a). $2mn - (m^2 - n^2) = 1$ can be rewritten $(m+n)^2 - 2m^2 = 1$, while $(m^2 - n^2) - 2mn = 1$ can be rewritten $(m-n)^2 - 2n^2 = 1$. We obtain Pell's equation $x^2 - 2y^2 = 1$, where $(x, y) = (m+n, m)$, $(m, n) = (y, x-y)$, and $(x, y) = (m-n, n)$, $(m, n) = (x+y, y)$.

3.2(b). $(m, n) = (2, 1)$ gives the triple (3, 4, 5), while $(m, n) = (5, 2)$ gives the triple (21, 20, 29).

3.2(c). For example, $(x, y) = (17, 12)$ leads ultimately to the triples (119, 120, 169) and (697, 696, 985).

3.3. $(a+c)^2 + (a+c+1)^2 - (2a+c+1)^2 = 1 - \left[a^2 + (a+1)^2 - c^2\right]$.

3.4(a).

$$\begin{aligned} q_{n+1}^2 - q_n^2 - 2q_{n+1}q_n &= (q_{n+1} - 2q_n)q_{n+1} - q_n^2 = q_{n-1}q_{n+1} - q_n^2 \\ &= q_{n-1}(2q_n + q_{n-1}) - q_n(2q_{n-1} + q_{n-2}) \\ &= q_{n-1}^2 - q_n q_{n-2} \\ &= \cdots = (-1)^n(q_1^2 - q_0 q_2) = (-1)^n. \end{aligned}$$

3.5. $2mn - (m^2 - n^2) = 7$ can be rewritten $(n+m)^2 - 2m^2 = 7$, and $(m^2 - n^2) - 2mn = 7$ as $(m-n)^2 - 2n^2 = 7$. The equation $x^2 - 2y^2 = 7$ is satisfied by $(|x|, |y|)] = (3, 1), (13, 9), (75, 53)$, and these yield the triples (−3, 4, 5), (8, 15, 17), (65, 72, 97), (396, 403, 565), (2325, 2332, 3293), (13568, 13575, 19193).

3.6. Let $a^2 + b^2 = c^2$, $p^2 + q^2 = r^2$ with $p = a+3$, $q = b+3$, and $r = c+4$. Then we must have $3a + 3b + 1 = 4c$. Taking $(a, b, c) = (m^2 - n^2, 2mn, m^2 + n^2)$, we are led to $(m - 3n)^2 - 2n^2 = 1$. Suppose that $x = m - 3n$ and $y = n$. Then $x^2 - 2y^2 = 1$. Some solutions are

$$(x, y) = (3, 2), (-3, 2), (17, 12), (-17, 12),$$

which give rise to

$$(m, n) = (9, 2), (3, 2), (53, 12), (19, 12).$$

Thus we have some examples:

$$\begin{aligned} [(a, b, c), (p, q, r)] = {} & [(77, 36, 85), (80, 39, 89)], [(5, 12, 13), (8, 15, 17)], \\ & [(2665, 1272, 2953), (2668, 1275, 2957)], \\ & [(217, 456, 505), (220, 459, 509)]. \end{aligned}$$

Comments on the Explorations

Exploration 1.2. If $x^2 - 2y^2 = \pm 2$, then x would have to be even, say $x = 2z$, so that $y^2 - 2z^2 = \mp 1$. The equation $x^2 - 2y^2 = 2$ is satisfied by $(x, y) = (2, 1), (10, 7), (58, 41)$, and $x^2 - 2y^2 = -2$ by $(x, y) = (4, 3), (24, 17), (140, 99)$. As for $x^2 - 2y^2 = 3$, if y is even, then $x^2 \equiv 3 \pmod 8$, while if y is odd, then $x^2 \equiv 5 \pmod 8$; neither congruence for x^2 is realizable. A similar argument shows the impossibility of $x^2 - 2y^2 \equiv -3$.

Exploration 1.3. We have to solve the equation $x^2 - 3y^2 = 1$, where $x = 2u$ is even and $y = 2v + 1$ is odd. The following table gives the first few triangles of the required type:

(x, y)	$(v, v+1, u)$
(26, 15)	(7, 8, 13)
(362, 209)	(104, 105, 181)
(5042, 2911)	(1455, 1456, 2521)
(70226, 40545)	(20272, 20273, 35113)
(978122, 564719)	(282359, 282360, 489061)

Observe that the equation satisfied by u and v implies that $(v + 1)^3 - v^3 = u^2$. It is known from this that u must be the sum of two consecutive squares. (See *Power Play* by E.J. Barbeau (MAA, 1997) pages 7, 8, 11 for a proof of this.)

$$\begin{aligned} 13 &= 2^2 + 3^2, \\ 181 &= 9^2 + 10^2, \\ 2521 &= 35^2 + 36^2, \\ 35113 &= 132^2 + 133^2, \\ 489061 &= 494^2 + 495^2. \end{aligned}$$

Adding together the consecutive roots of the squares in each right side gives the sequence $\{5, 19, 71, 265, 981, \ldots\}$, which satisfies the recursion $t_{n+1} = 4t_n - t_{n-1}$. The terms of the sequence are sums of the consecutive terms of a second sequence $\{1, 4, 15, 56, 209, \ldots\}$ that satisfies the same recursion. The values of y when $x^2 - 3y^2 = 1$ are not only found among its terms, but the numbers x and y are intricately related in other ways. For example, $(5042, 2911) = (71^2 + 1, 56^2 - 15)$. Is there something here that can be generalized?

Exploration 1.4. This exploration can be approached in various ways. For $x^2 + mx \pm n$ to be factorizable, the discriminants $m^2 \mp 4n$ must be integer squares, say $m^2 - 4n = u^2$ and $m^2 + 4n = v^2$. In particular, we need $u^2 + v^2 = 2m^2$. We can achieve this if $m^2 = p^2 + q^2$. Then $2m^2 = (p + q)^2 + (p - q)^2$, so we can try $u = p - q$ and $v = p + q$. This leads to $n = pq/2$. Thus we can find polynomials involving any number m whose square is the sum of two squares. Since m is thus the largest of a Pythagorean triple, we can let $m = \gamma(\alpha^2 + \beta^2)$,

$p = \gamma(\alpha^2 - \beta^2)$, and $q = 2\gamma\alpha\beta$. Then $n = pq/2 = \gamma^2\alpha\beta(\alpha^2 - \beta^2)$. This leads to

$$x^2 + mx + n = (x + \gamma\alpha(\alpha - \beta))(x + \gamma\beta(\alpha + \beta)),$$
$$x^2 + mx - n = (x + \gamma\alpha(\alpha + \beta))(x - \gamma\beta(\alpha - \beta)).$$

Alternatively, we need to find numbers a, b, c, d for which $(x + a)(x + b) = (x - c)(y + d)$. This yields $m = a + b = d - c$ and $n = ab = cd$. Substituting $d = ab/c$ into the first equation gives $c^2 + (a + b)c - ab = 0$ with discriminant $(a + 3b)^2 - 8b^2$. If this is the square of w, then

$$8b^2 = (a + 3b)^2 - w^2 = (a + 3b + w)(a + 3b - w).$$

To construct possibilities, let b be chosen arbitrarily and write $8b^2 = (2r)(2s)$. Then solving $a + 3b + w = 2r$, $a + 3b - w = 2s$ gives $a + 3b = r + s$ and $w = r - s$, leading to the possibilities

$$(a, b, c, d) = (r + s - 3b, b, b - s, r - b),$$
$$(m, n) = (r + s - 2b, b(r + s) - 3b^2).$$

Thus

$$x^2 + mx + n = (x + b)(x + r + s - 3b)$$

and

$$x^2 + mx - n = (x + s - b)(x + r - b).$$

Pell's equation is at the heart of a third approach. With $m^2 - 4n = u^2$ and $m^2 + 4n = v^2$, we obtain $v^2 - 2m^2 = -u^2$. We could take, for example, $u = 1$. A possible solution is $(u, v, m) = (1, 41, 29)$, leading to $n = 210$ and

$$x^2 + 29x + 210 = (x + 15)(x + 14) \quad \text{and} \quad x^2 + 29x - 210 = (x + 35)(x - 6).$$

Exploration 1.5. Note the following chain of equivalent equations:

$$[k(m^2 - n^2) + a]^2 + [2kmn + a]^2 = [k(m^2 + n^2) + (a + 1)]^2,$$
$$a^2 - 2a - 1 = 2km^2 - 4akmn + (4ak + 2k)n^2,$$
$$(a - 1)^2 - 2 = 2k[(m - an)^2 - (a^2 - 2a - 1)n^2],$$
$$\frac{(a - 1)^2 - 2}{2k} = (m - an)^2 - (a^2 - 2a - 1)n^2.$$

If $k = 1$, we get

$$\frac{(a - 1)^2}{2} - 1 = (m - an)^2 - ((a - 1)^2 - 2)n^2.$$

For example, if $a = 11$, we are led to $49 = (m - 11n)^2 - 98n^2$. Thus, $(m, n) = (7, 0)$ leads to the pair $\{(0, 49, 49), (11, 60, 61)\}$, and $(m, n) = (77, 70)$ to $\{(1029, 10780, 10829), (1040, 10791, 10841)\}$.

Suppose that $a = 2b + 1$ and $k = 2b^2 - 1$. Then we get the equation

$$1 = (m - an)^2 - (4b^2 - 2)n^2,$$

which is satisfied by $(m, n) = (2b + 1, 2b)$ and $(8b^2 + 2b - 1, 2b)$. The first of these yields the two triples

$$(8b^3 + 2b^2 - 4b - 1, 16b^4 + 8b^3 - 8b^2 - 4b, 16b^4 + 8b^3 - 6b^2 - 4b - 1),$$
$$(8b^3 + 2b^2 - 2b, 16b^4 + 8b^3 - 8b^2 - 2b + 1, 16b^4 + 8b^3 - 6b^2 - 2b + 1).$$

Chapter 2

1.1. For 1, $(x, y, m, n) = (3, 1, 1, 1)$, and for 36, $(x, y, m, n) = (17, 6, 6, 8)$. Another possibility is $(x, y, m, n) = (99, 35, 35, 49)$, yielding the square and triangular number $1225 = 35^2 = 1 + 2 + \cdots + 49$. The values of n are, in fact $1, 8 = 3^2 - 1, 49 = 7^2, 288 = 17^2 - 1, 1681 = 41^2, \ldots$.

1.2. $\frac{1}{2}n(n + 1) + 1 = m^2$ can be rewritten $(2n + 1)^2 - 8m^2 = -7$. Some solutions are $(m, n) = (1, 0), (2, 2), (4, 5), (11, 15), (23, 32), (64, 90)$. Also, $\frac{1}{2}n(n + 1) - 1 = m^2$ can be rewritten as $(2n + 1)^2 - 8m^2 = 9$. Some soutions are $(m, n) = (0, 1), (3, 4), (18, 25), (105, 148)$.

1.3. The product of $\frac{1}{2}(n - 1)n$, $\frac{1}{2}n(n + 1)$ and $\frac{1}{2}(n + 1)(n + 2)$ is a square if and only if $\frac{1}{2}(n - 1)(n + 2) = m^2$ for some integer m. This can be rewritten as $(2n + 1)^2 - 8m^2 = 9$. Some solutions are $(m, n) = (0, 1), (3, 4), (18, 25), (105, 148), (612, 865)$.

1.4. The sum of $\frac{1}{2}(n - 1)n$, $\frac{1}{2}n(n + 1)$ and $\frac{1}{2}(n + 1)(n + 2)$ is a square if and only if $3n^2 + 3n + 2 = 2m^2$ for some integer m. This can be rewritten as $(4m)^2 - 6(2n + 1)^2 = 10$. Some solutions are $(m, n) = (2, 1), (8, 6), (19, 15), (79, 64)$.

1.5. The condition is that $n(n+1) = 2m(m+1)$ or $(2n+1)^2 - 2(2m+1)^2 = -1$. Some solutions are $(m, n) = (2, 3), (14, 20), (84, 119)$.

1.6. We obtain $n(n + 1) - m(m - 1) = 2mn$, which, after multiplication by 4 and some manipulation, leads to $p^2 - 8n^2 = 1$ with $p = 2m + 2n - 1$. Some solutions are

$$(p, m, n) = (3, 1, 1), (17, 3, 6), (99, 15, 35), (577, 85, 204), (3363, 493, 1189).$$

If (m_k, n_k) is the kth solution, then

$$(m_{k+1}, n_{k+1}) = (6m_k - m_{k-1} - 2, 6n_k - n_{k-1}) = (m_k + 2n_k, 2m_k + 5n_k - 1).$$

2.1. The condition on a and b is $a(b + 1) = (a - b)(a - b - 1)$, which reduces to $0 = a^2 - 3ab + b^2 - 2a + b$. We set this up for completing the square. Multiplying the equation by 4 leads to

$$0 = (2a - 3b - 2)^2 - (5b^2 + 8b + 4).$$

To complete the square for the second term on the right, multiply the equation by 5 to obtain

$$0 = 5(2a - 3b - 2)^2 - (25b^2 + 40b + 20) = 5(2a - 3b - 2)^2 - (5b + 4)^2 - 4.$$

Thus, we need to solve $x^2 - 5y^2 = -4$, where $x = 5b + 4$, $y = 2a - 3b - 2$. Two solutions are $(x, y) = (29, 13)$ and $(9349, 4181)$, which lead to $(a, b) = (15, 5)$ and $(4895, 1869)$.

2.2. Of the $\binom{n}{2}$ ways of drawing a pair of marbles, $r(n - r)$ will produce marbles of different colors. We require $n(n - 1) = 4r(n - r)$. While this could be cast as a Pell's equation, in this case it is more convenient to render it in the form $n = (n - 2r)^2$. Then $n = m^2$ for some integer m, so that $m^2 - 2r = \pm m$. Thus

$$r = \frac{m^2 \mp m}{2} = \binom{m}{2} \quad \text{or} \quad \binom{m+1}{2}.$$

Each value of m actually yields a solution, so

$$(n, r) = \left(m^2, \binom{m}{2}\right) \quad \text{and} \quad \left(m^2, \binom{m+1}{2}\right)$$

covers all cases.

2.3(a). Suppose $a + 1 = y^2$ and $3a + 1 = x^2$. Then $x^2 - 3y^2 = -2$, which is satisfied by $(x, y) = (1, 1)$. If $(x, y) = (u, v)$ is a solution, then so is $(x, y) = (2u + 3v, u + 2v)$. This leads to a succession of solutions

$$(x, y) = (1, 1), (5, 3), (19, 11), (71, 41), (265, 153), (989, 571), \ldots$$

whose corresponding values of a are 0, 8, 120, 1680, 23408, 326040, ... This list extends to include all possibilities.

2.3(b) Let $u_0 = v_0 = 1$, $a_0 = 0$, and for $n \geq 0$, let

$$u_{n+1} = 2u_n + 3v_n, \quad v_{n+1} = u_n + 2v_n, \quad a_n = v_n^2 - 1.$$

Playing around, we note that

$$\begin{aligned} 8 \times 5 \times 3 &= 120 - 0, \\ 8 \times 19 \times 11 &= 1680 - 8, \\ 8 \times 71 \times 41 &= 23408 - 120, \ldots, \\ 8 \times 120 + 1 &= 31^2, \\ 120 \times 1680 + 1 &= 449^2, \\ 1680 \times 23408 + 1 &= 6271^2, \ldots, \end{aligned}$$

and

$$31 = 2 \times 3 \times 5 + 1, \quad 449 = 2 \times 11 \times 19 + 31, \quad 6271 = 2 \times 42 \times 71 + 449, \ldots.$$

This leads to a number of conjectures:

$$a_{n+1} - a_{n-1} = 8u_n v_n, \tag{1}$$

$$a_n a_{n+1} + 1 = [2(u_n v_n + u_{n-1} v_{n-1} + \cdots + u_1 v_1) + 1]^2, \tag{2}$$

for $n \geq 1$. If (1) is true, then (2) is equivalent to

$$8u_n v_n a_n + (a_n a_{n-1} + 1) = 4u_n^2 v_n^2 + 4u_n v_n (2u_{n-1}v_{n-1} + \cdots + 2u_1 v_1 + 1) + [2(u_{n-1}v_{n-1} + \cdots + u_1 v_1) + 1]^2$$

or

$$2a_n = u_n v_n + 2u_{n-1}v_{n-1} + \cdots + 2u_1 v_1 + 1 \tag{3}$$

for $n \geq 1$. If (1) is true, then (3) is equivalent to

$$u_{n+1}v_{n+1} - 14u_n v_n + u_{n-1}v_{n-1} = 0 \tag{4}$$

for $n \geq 1$.

It suffices then to establish (1) and (4). Since

$$(u_{n-1}, u_n, u_{n+1}) = (u_{n-1}, 2u_{n-1} + 3v_{n-1}, 7u_{n-1} + 12v_{n-1})$$

and

$$(v_{n-1}, v_n, v_{n+1}) = (v_{n-1}, u_{n-1} + 2v_{n-1}, 4u_{n-1} + 7v_{n-1}),$$

then

$$(u_{n-1}v_{n-1}, u_n v_n, u_{n+1}v_{n+1}) = (u_{n-1}v_{n-1}, 2u_{n-1}^2 + 7u_{n-1}v_{n-1} + 6v_{n-1}^2, 28u_{n-1}^2 + 97u_{n-1}v_{n-1} + 84v_{n-1}^2),$$

whence (4) holds.

In a similar way, we establish that $v_{n+1} = 4v_n - v_{n-1}$, whence

$$\begin{aligned} a_{n+1} - a_{n-1} &= (v_{n+1}^2 - v_{n-1}^2) = (v_{n+1} + v_{n-1})(v_{n+1} - v_{n-1}) \\ &= 4v_n(v_{n+1} - v_{n-1}) = 4v_n(4u_{n-1} + 6v_{n-1}) \\ &= 8v_n(2u_{n-1} + 3v_{n-1}) = 8v_n u_n, \end{aligned}$$

as desired.

Combining (2) and (1) leads to

$$16(a_n a_{n+1} + 1) = (a_{n+1} + a_n - 4)^2.$$

Since $a_{n+1} + a_n = 8b_n^2$, we have that $a_n a_{n+1} + 1 = (2b_n^2 - 1)^2$, where $b_0 = 1$, $b_1 = 4$, $b_{n+1} = 4b_n - b_{n-1}$ for $n \geq 1$. The reader may wish to establish this, as well as $a_{n+2} = 14a_{n+1} - a_n + 8$ for $n \geq 1$.

2.4. In particular, we require that $1 + b = \frac{1}{2}u(u+1)$ and $1 + b + b^2 = \frac{1}{2}v(v+1)$ for some integers u and v. The second equation can be rewritten in the form $x^2 - 2y^2 = 7$ with $x = 2v + 1$ and $y = 2b + 1$. These have solutions

$$(x, y; v, b) = (3, 1; 1, 0), (5, 3; 2, 1), (13, 9; 6, 4), (27, 19; 13, 9), (75, 53; 37, 26), (157, 111; 78, 55), \ldots.$$

Of the bases 4, 9, 26, 55, only 9 is one less than a triangular number. In fact, for any number k of digits, the integer

$$(111\ldots 1)_9 = 1 + 9 + \cdots + 9^{k-1} = \frac{9^k - 1}{8} = \frac{1}{2}\left[\left(\frac{3^k - 1}{2}\right)\left(\frac{3^k + 1}{2}\right)\right]$$

is a triangular number. Whether any other bases b exist is an open question.

2.5. $\sum_{k=1}^{n}(2k)^2 = 2n(n+1)(2n+1)/3$, while

$$\begin{aligned}\sum_{r=1}^{n}(m+r-1)(m+r) &= \frac{1}{3}\sum_{r=1}^{n}[(m+r-1)(m+r)(m+r+1) \\ &\qquad - (m+r-2)(m+r-1)(m+r)] \\ &= \frac{(m+n-1)(m+n)(m+n+1)-(m-1)m(m+1)}{3} \\ &= nm^2+n^2m+\frac{n^3-n}{3}.\end{aligned}$$

Equating the two leads to $(n+1)^2 = m(n+m)$, or $n^2+[2(n+1)]^2 = (2m+n)^2$. From Exercise 1.3.1 we can determine positive integers z and y for which $n = y^2 - z^2$ and $n+1 = yz$. This leads to $z^2+zy-y^2 = 1$, or $x^2-5y^2 = 4$, where $x = 2z+y$.

$$\begin{aligned}(x, y) &= (3,1), (7,3), (18,8), (47,21), (123,55), (322,144), (843,377), \\ (m, n) &= (1,0), (4,5), (25,39), (169,272), (1156,1869), \\ &\qquad (7921, 12815), (54289, 87840).\end{aligned}$$

We note in passing the intervention of the Fibonacci sequence, with $F_0 = 0$, $F_1 = 1$ and $F_{n+1} = F_n + F_{n-1}$ for $n \geq 1$. The reader is invited to show that $(y, z) = (F_{2k}, F_{2k-1})$, whereupon $(m, n) = (F_{2k-1}^2, F_{2k-2}F_{2k+1})$.

2.6. Subtracting twice the square of the second equation from the square of the first equation yields

$$-32 = 4u^2v^2 - 2x^2v^2 - 2u^2y^2 + x^2y^2 = (x^2-2u^2)(y^2-2v^2).$$

Solving equations of the form $x^2 - 2u^2 = \pm 2^r$, $y^2 - 2v^2 = \mp 2^s$, where r and s are nonnegative integers summing to 5, and checking for extraneous solutions leads to the result.

For example, the system $x^2 - 2u^2 = 2$, $y^2 - 2v^2 = -16$ along with the given system is satisfied by

$$(x, y, u, v) = (2, -4, 1, 4), (10, 4, 7, 4), (2, -28, -1, 20).$$

Note that $(|x|, |y|, |u|, |v|) = (10, 28, 7, 20)$ also satisfies the pair of Pell's equations, but no adjustment of signs of the entries will yield a solution of the original system.

The pair $x^2 - 2u^2 = \pm 2^r$, $y^2 - 2v^2 = \mp 2^s$ will arise from any system $2uv - xy = a, xv - uy = b$ for which $a^2 - 2b^2 = -32$. Each of a and b must be divisible by 4; possible values are given by $(|a|, |b|) = (0, 4), (16, 12), (96, 68)$.

2.7. $2x^2 + (x \pm 1)^2 = w^2$ can be rewritten as $(3x \pm 1)^2 - 3w^2 = -2$. Some solutions are $(3x \pm 1, w) = (1, 1), (5, 3), (19, 11), (71, 41), (265, 153)$. Those corresponding to integer values of x give the quadruples

$$(x, y, z, w) = (0, 0, 1, 1), (2, 2, 1, 3), (6, 6, 7, 11), (24, 24, 23, 41), (88, 88, 89, 153).$$

There are infinitely many others.

2.8. The condition on n is that $(n + 1)(2n + 1) = 6m^2$ for some integer m. This can be rewritten as $(4n + 3)^2 - 3(4m)^2 = 1$. Thus, we wish to find solutions of $x^2 - 3y^2 = 1$ for which $x + 1$ and y are multiplies of 4. Two possibilities are $(x, y) = (7, 4), (1351, 780)$, leading to $n = 1$ and $n = 337$.

2.9. We can rewrite the equation in the form $z^2 - (1 + x^2)y^2 = x^2$, which has the obvious solutions $y = 0, z = x$ and $x = 0, z = y$. To find nonzero solutions, try various values of x. For example, $x = 1$ leads to $z^2 - 2y^2 = 1$, $x = 2$ to $z^2 - 5y^2 = 4$ and $(x, y, z) = (2, 1, 3), (2, 8, 18), (2, 21, 47)$, and $x = 3$ to $z^2 - 10y^2 = 9$ and $(x, y, z) = (3, 18, 57)$. It is interesting to note that each term of the triple $(2, 8, 18)$ is twice a square. There are solutions for each value of x.

2.10. Let $a = \frac{1}{2}(m^3 + m^2) - 1$, $b = \frac{1}{2}(m^3 - m^2) + 1$, $c = m^2$. Then $2s = m^3 + m^2$, so $s - a = 1$, $s - b = m^2 - 1$, $s - c = \frac{1}{2}(m^3 - m^2)$, and the area, Δ, is $\frac{1}{2}m^2(m^2 - 1)$. Again, let $a = m^3 - \frac{1}{2}(m - 1)$, $b = m^3 - \frac{1}{2}(m + 1)$, $c = m$. Then $s = m^3$, $s - a = \frac{1}{2}(m - 1)$, $s - b = \frac{1}{2}(m + 1)$, $s - c = m(m^2 - 1)$, so that $\Delta = \frac{1}{2}m^2(m^2 - 1)$.

Suppose that $\frac{1}{2}(m^2 - 1) = n^2$. Then $m^2 - 2n^2 = 1$. This is satisfied, for example, by $(m, n) = (3, 2), (17, 12)$, which yields the triangles

$$\begin{aligned}(a, b, c; s, \Delta) = {} & (17, 10, 9; 18, 36), (26, 25, 3; 27, 36),\\ & (2600, 2313, 289; 2601, 41616),\\ & (4905, 4904, 17; 4913, 41616).\end{aligned}$$

2.11(a). With $\Delta = \sqrt{s(s - a)(s - b)(s - c)}$, we are led to $16\Delta^2 = 3t^2(t^2 - 4)$, so that $3(t^2 - 4)$ is an even square. Setting $t = 2x$, we see that $4 \cdot 3(x^2 - 1)$ is a square. Hence $3(x^2 - 1)$ must be square, and, being divisible by 3, have the form $9y^2$. Hence $x^2 - 3y^2 = 1$. Some examples of triangles with integer areas are (3, 4, 5), (13, 14, 15), (51, 52, 53), (193, 194, 195), (723, 724, 725), (2701, 2702, 2703), (10083, 10084, 10085), (37633, 37634, 37635). See *College Mathematics Journal* 29 (1998), 13–17.

2.11(b). $\Delta = 3yt/2$, so that the length of the altitude is $3y$. The side of length t is partitioned into sides of lengths

$$\sqrt{(2x - 1)^2 - (3y)^2} = \sqrt{(4x^2 - 4x + 1) - 3(x^2 - 1)} = \sqrt{x^2 - 4x + 4} = x - 2$$

and

$$\sqrt{(2x + 1)^2 - (3y)^2} = \sqrt{(4x^2 + 4x + 1) - 3(x^2 - 1)} = \sqrt{x^2 + 4x + 4} = x + 2.$$

2.12(a). By Pythagoras' theorem

$$(ma + r)^2 - [(ma - r)^2 - q^2] = (a \pm q)^2,$$

which reduces to $4mar = a^2 \pm 2aq$.

2.12(b). $4c^2 = (2p)^2 + (2q)^2$, so that $(2p)^2$ is an integer. Hence $2p$ must be an integer.

2.12(c). With $s = \frac{1}{2}(a + b + c) = ma + \frac{a}{2}$, we obtain

$$\left(\frac{ap}{2}\right)^2 = a\left(m + \frac{1}{2}\right)a\left(m - \frac{1}{2}\right)\left(\frac{a}{2} + r\right)\left(\frac{a}{2} - r\right),$$

which leads to the desired equation.

2.12(d). An obvious solution to the equation in (c) is $(p, t) = (0, r)$. Other solutions are

$$(p, t) = ((4m^2 - 1)r, 2mr), \quad (4mr(4m^2 - 1), (8m^2 - 1)r).$$

This yields triangles

$$\begin{aligned}(a, b, c) = (4mr, (4m^2 + 1)r, (4m^2 - 1)r),\\ (2(8m^2 - 1)r, (16m^3 - 2m + 1)r, (16m^3 - 2m - 1)r).\end{aligned}$$

Other solutions appear. For example, $(m, r) = (4, 3)$ leads to $p^2 - 63t^2 = -567$, for which one solution is $(p, t) = (21, 4)$ leading to the triangle $(a, b, c) = (8, 35, 29)$.

2.13. Solving $n^2 - 2m^2 = 1$ leads to the triples (8, 9, 10), (288, 289, 290), (9800, 9801, 9802), for example. Here are some other approaches:

(i) $(u^2, u^2 + 1, 2v^2)$, where $u^2 - 2v^2 = -2$, leading to (16, 17, 18), (576, 577, 578).

(ii) $(4r^2(r^2 + 1), (2r^2 + 1)^2, 4r^4 + 4r^2 + 2)$, yielding (8, 9, 10), (80, 81, 82), (360, 361, 362).

(iii) $(n^2, n^2 + 1, (n - 1)^2 + m^2)$, where $m^2 = 2n + 1$, yielding (16, 17, 18), (144, 145, 146), (576, 577, 578).

Since no number congruent to 3, modulo 4, can be written as the sum of two squares, triples of the required type must begin with a multiple of 4. Furthermore, a number can be expressed as the sum of two squares if and only if no prime congruent to 3, modulo 4, divides it to an odd power.

3.7. $(1728148040)^2 - 151(140634693)^2 = 1$ is equivalent to

$$151(140634693)^2 = (1728148041)(1728148039).$$

Checking for divisibility by 151 yields $1728148039 = 151 \times 11444689$. Checking for divisibility by 9, we find that $140634693 = 9 \times 15626077$ and $1728148041 = 9^2 \times 21335161$. Checking for other small prime divisors gives $15626077 = 17 \times 919181$ and $11444689 = 17^2 \times 39601$. To get a handle on further divisions, we take the greatest common divisor of 919181 and

39601: gcd(919181, 39601)=gcd(39601, 8358)=gcd(8358, 6169) = gcd(6169, 2189)=gcd(2189, 1791)=gcd(1791, 398)=gcd(398, 199) = 199, where gcd(x, y) is the greatest common divisor of x and y. Accordingly,

$$151 \times 140634693^2 = 151 \times (9 \times 17 \times 199 \times 4619)^2$$

and

$$1728148041 \times 1728148039 = 9^2 \times 21335161 \times 151 \times 17^2 \times 199^2.$$

Since $21335161 = 4619^2$, we have finished our checking.

Since $12^2 + 13^2 = 313$, $1268622368^2 - 313 \times 7170685^2 = -1$ is equivalent to

$$126862368^2 - 12^2 \times 7170685^2 = 13^2 \times 7170685^2 - 1,$$

or

$$126862368^2 - 86048220^2 = 93218905^2 - 1,$$

or

$$212910588 \times 40814148 = 93218906 \times 93218904.$$

Now

$$\begin{aligned} 212910588 &= 2^2 \times 3^2 \times 11 \times 537653, \\ 40814148 &= 2^2 \times 3 \times 3401179, \\ 93218906 &= 2 \times 11 \times 4237223, \\ 93218904 &= 2^3 \times 3^3 \times 431569. \end{aligned}$$

The greatest common divisor of 537653 and 431569 is 2411, and we find that $537653 = 2411 \times 223$ and $431569 = 2411 \times 179$. Now verify that $4237223 = 223 \times 19001$ and $3401179 = 179 \times 19001$ to finish the checking.

3.10(b). $1 \leq y \leq 2z \Rightarrow z^2 + 1 \leq z^2 + (2z/y) \leq z^2 + 2z$, while $-2z \leq y \leq -1 \Rightarrow z^2 - 2z \leq z^2 + (2z/y) \leq z^2 - 1$.

3.10(c). $d = 3$ can be covered by $(y, z) = (1, 1), (-4, 2), (-1, 3)$, yielding solutions $(x, y) = (2, 1), (-7, -4), (-2, -1)$. $d = 27$ is covered by $(y, z) = (5, 5)$, yielding the solution $(x, y) = (26, 5)$. $d = 35$ is covered by $(y, z) = (1, 5), (-12, 6), (-1, 7)$, yielding solutions $(x, y) = (6, 1), (-71, -12), (-6, -1)$. As for $d = 45$, we cannot have $45 = z^2 + (2z/y)$ for integers y and z, but we note that $45 = 3^2 \times 5$. Accordingly, we find (x, y) such that $x^2 - 5y^2 = 1$ and y is divisible by 3. The method gives $(x, y) = (9, 4)$, but use of the identity in Exercise 3.4(a) leads to the solution $(x, y) = (161, 72)$. We find that $(x, y) = (161, 24)$ satisfies $x^2 - 45y^2 = 1$.

3.10(d). The method does not work for $d = 13, 19, 21, 22, 28, 29, 31, 41, 43, 44, 45, 46$. However, $28 = 2^2 \times 7$ and $45 = 3^2 \times 5$, so we can get solutions in these cases from the solutions for $d = 7$ and $d = 5$ respectively.

Comments on the Explorations

Exploration 2.1. The sum of the first n terms is $2n^2 - n$, and this is square if and only if $(4n-1)^2 - 8m^2 = 1$ for some integer m. The smallest values of n are $1, 25, 841, 28561$. These are all squares (of $1, 5, 29, 169$) and appear to be squares of terms q_k for odd indices k.

Exploration 2.2. We are looking for triples (k, m, n) for which $k(k+1) \cdot m(m+1) = n(n+1)$. This is equivalent to

$$x^2 - k(k+1)y^2 = -k^2 - k + 1, \tag{1}$$

where $x = 2n+1$ and $y = 2m+1$. Take k as a parameter. Equation (1) has two obvious solutions: $(x, y) = (1, 1), (-1, 1)$. From the fact that

$$x^2 - k(k+1)y^2 = 1 \tag{2}$$

is satisfied by $(x, y) = (2k+1, 2)$, we deduce that if $(x, y) = (u, v)$ satisfies (1), then so also does $(x, y) = ((2k+1)u + 2k(k+1)v, 2u + (2k+1)v)$. Thus $(x, y) = (1, 1)$ gives rise to the solutions

$$(x, y) = (1, 1), (2k^2+4k+1, 2k+3), (8k^3+20k^2+12k+1, 8k^2+16k+5), \ldots,$$

while $(x, y) = (-1, 1)$ yields

$$(x, y) = (-1, 1), (2k^2-1, 2k-1), (8k^3+4k^2-4k-1, 8k^2-3), \ldots.$$

Using this, we can construct infinitely many oblong triples corresponding to each value of k:

$$\begin{aligned}(k, m_i, n_i) &= (k, 0, 0), (k, k+1, k^2+2k), \\ &\qquad (k, 4k^2+8k+2, 4k^3+10k^2+6k), \ldots, \\ &= (k, 0, -1), (k, k-1, k^2-1), \\ &\qquad (k, 4k^2-2, 4k^3+2k^2-2k-1), \ldots.\end{aligned}$$

The sequences of $\{m_i\}$ and $\{n_i\}$ satisfy the recursion

$$t_n = (4k+2)t_{n-1} - t_{n-2} + 2k.$$

The solutions of the equations (1) appear to have some interesting linkages. For example,

$$\begin{aligned} x^2 - 2y^2 &= -1 \quad \text{is satisfied by} \quad (x, y) = (7, 5), (239, 169); \\ x^2 - 6y^2 &= -5 \quad \text{is satisfied by} \quad (x, y) = (17, 7), (169, 69); \\ x^2 - 12y^2 &= -11 \quad \text{is satisfied by} \quad (x, y) = (17, 5), (239, 69). \end{aligned}$$

There may be other instances of this sort of thing.

Exploration 2.3. The relationship involving the Fibonacci numbers can be established using

$$\begin{aligned} F_m^2 - F_{m+1}F_{m-1} &= F_m^2 - (F_m + F_{m-1})F_{m-1} = F_m(F_m - F_{m-1}) - F_{m-1}^2 \\ &= -[F_{m-1}^2 - F_mF_{m-2}] \end{aligned}$$

and

$$\begin{aligned} F_{2n+1}^2 + F_{2n-1}^2 + 1 &= (F_{2n+1} - F_{2n-1})^2 + 2F_{2n-1}F_{2n+1} + 1 \\ &= F_{2n}^2 + 1 + 2F_{2n-1}F_{2n+1}. \end{aligned}$$

With respect to

$$(2a - kb)^2 - (k^2 - 4)b^2 = -4,$$

if $k \not\equiv 0 \pmod 3$, then $(2a - kb)^2 \equiv -4 \equiv 2 \pmod 3$, an impossibility. If $k = 2l$ with l odd, then

$$0 = (a - lb)^2 - (l^2 - 1)b^2 + 1 \equiv (a - lb)^2 + 1 \pmod 8,$$

again an impossibility.

If k is even with $k = 4l$ for some integer l, the equation becomes

$$(a - 2lb)^2 - (4l^2 - 1)b^2 = -1.$$

Now, b cannot be even, since then $(a - 2lb)^2 \equiv 1 \pmod 4$, an impossibility. If b were odd, then $(a - 2lb)^2 + b^2 \equiv 3 \pmod 4$, again an impossibility. When k is odd, the equation $x^2 - (k^2 - 4)y^2 = 4$ has smallest solution in positive integers $(x, y) = (k, 1)$. As we shall see in Chapter 5, this precludes a solution of $x^2 - (k^2 - 4)y^2 = -4$, except for the case $k = 3$.

Exploration 2.4. This is Problem 10622 in *American Mathematical Monthly* [1997, 870; 1999, 867–868]. Other triples along with the squares involved are (0, 2, 4; 1, 1, 3), (1, 8, 15; 3, 4, 11), (4, 30, 56; 11, 15, 41). This can be generalized by the sets

$$(a_n, 2a_{n+1}, a_{n+2}; a_{n+1} - a_n, a_{n+1}, 3a_{n+1} - a_n),$$

where $a_0 = 1, a_1 = 4, a_{n+2} = 4a_{n+1} - a_n$. Alternatively, we can look at solutions to the Pell's equation $x^2 - 3y^2 = 1$. If $x_n + y_n\sqrt{3} = (2 + \sqrt{3})^n$, then possibilities are given by

$$(2y_n - x_n, 2y_n, 2y_n + x_n; x_n - y_n, y_n, x_n + y_n).$$

Exploration 2.5. $(n - 1)^2 + n^2 + (n + 1)^2 = k[(m - 1)^2 + m^2 + (m + 1)^2]$ leads to $3(n^2 - km^2) = 2(k - 1)$. When $k = 3u + 1$, the equation becomes $n^2 - km^2 = 2u$. When $u = 2$, we get $n^2 - 7m^2 = 4$ with solutions $(n, m) = (2, 0), (16, 6), (254, 96), \ldots$. When $u = 3$, we have $n^2 - 10m^2 = 6$ with solutions $(n, m) = (4, 1), (136, 43), \ldots$.

For

$$(n-2)^2 + (n-1)^2 + n^2 + (n+1)^2 + (n+2)^2$$
$$= 2[(m-2)^2 + (m-1)^2 + m^2 + (m+1)^2 + (m+2)^2],$$

we have to solve $n^2 - 2m^2 = 2$ and get solutions $(n, m) = (2, 1), (10, 7), (58, 41), \dots$.

Exploration 2.8. This discussion involves material in later chapters, and you should defer reading it until you finish chapter 5. Suppose that p is a prime divisor of d. Then for any solution of the Pell's equation, we must have $x^2 \equiv -1 \pmod{p}$. It is known that this congruence is solvable if and only if $p = 2$ or $p \equiv 1 \pmod 4$. (See Exploration 4.1.) Thus, it is necessary that every odd prime divisor of d leave remainder 1 upon division by 4.

If $x^2 - dy^2 = -1$ has a solution with x even, then $dy^2 \equiv 1 \pmod 4$, so that y must be odd and $d \equiv 1 \pmod 4$. If there is a solution with x odd, then $dy^2 \equiv 2 \pmod 8$, so y must be odd and $d \equiv 2 \pmod 8$. Thus it is necessary that $d \equiv 1, 2, 5 \pmod 8$. Some values of d can be settled right away. If $d = k^2 + 1$, $(x, y) = (k, 1)$ works. For some values of d we can obtain solutions from odd solutions of $x^2 - dy^2 = -4$; it is necessary that $d \equiv 5 \pmod 8$. Suppose, for example, that d exceeds an odd square by 4, say $d = (2k+1)^2 + 4 = 4k^2 + 4k + 5$. Then we obtain from

$$\frac{1}{8}\left[(2k+1) + \sqrt{d}\right]^3$$

the rather intriguing solution

$$(x, y) = (4k^3 + 6k^2 + 6k + 2, 2k^2 + 2k + 1) = (2[k^3 + (k+1)^3], k^2 + (k+1)^2).$$

(Why does this not work when d exceeds an even square by 4?) On the other hand, suppose that $d = k^2 - 2$. Then $(x, y) = (k^2 - 1, k)$ satisfies $x^2 - dy^2 = 1$. If $x^2 - dy^2 = -1$ had a solution, then there would have to be one with $1 \le y \le k$. This is true for $k = 2$. Let $k > 2$. We first argue that $(k^2 - 1) + k\sqrt{d}$ cannot be the square of $a + b\sqrt{d}$, where $a, b \ge 1$. For otherwise, $a^2 + b^2(k^2 - 2) = k^2 - 1$, so $(b^2 - 1)k^2 + (a^2 + 1 - 2b^2) = 0$. Since $a^2 \pm 1 = db^2$, $a^2 + 1 > 2b^2$, and we arrive at a contradiction. Hence $(k^2 - 1) + k\sqrt{d}$ is not the square of an irrational $a + b\sqrt{d}$ whose norm is ± 1, and it follows that $x^2 - dy^2 = -1$ has no solution. This eliminates for example $d = 34$ and $d = 194$, not otherwise rejected.

Is there a solution whenever p is an odd prime that exceeds a multiple of 4 by 1?

Another way to look at the problem is to ask what possible values of y can occur. First of all, y must be odd, and secondly, y must be divisible only by primes congruent to 1 modulo 4. Thus y can only be 1, 5, 13, 17, 29. This picks up classes of d for which a solution occurs. For example, when $d = 25u^2 \pm 14u + 2$, $(x, y) = (25u \pm 7, 5)$, and when $d = 169u^2 \pm 140u + 29$, $(x, y) = (169u \pm 70, 13)$.

The issue of which values of d admit a solution of $x^2 - dy^2 = -1$ is deep and complex. For an exposition, see Chapter 3 of *Binary Quadratic Forms* by Duncan Buelle.

Exploration 2.10. If the first term of the arithmetic progression is 1, then there is only one possibility for each partial sum to be a square. Let $x_1 = 1$ and $x_2 = a^2 - 1$. Then $x_n = (n-1)a^2 - (2n-3)$ and

$$x_1 + x_2 + \cdots + x_n = \frac{1}{2}n(n-1)a^2 - (n-1)^2 + 1.$$

When $n = 9$, the partial sum is equal to $36a^2 - 63 = 9(4a^2 - 7)$. This is square if and only if $4a^2 - 7 = b^2$ or $7 = (2a - b)(2a + b)$ for some integer b. The only possibility is $2a - b = 1, 2a + b = 7$, where $a = 2$.

No partial sum of the arithmetic series $2 + 4 + 6 + \cdots$ is square, since each is the product of two consecutive positive integers. To get an example for which the initial term and common difference have greatest common divisor equal to 1, simply remove 1 from the progression of odd integers to get $3 + 5 + 7 + \cdots$.

Exploration 2.11. Let $(x_0, y_0) = (1, 0)$ and let (x_n, y_n) be the nth solution of $x^2 - dy^2 = 1$ when $d = 3$ or $d = 6$. Then

$$x_{2n-1} = (y_{n-1} + y_n)^2 + 1, \quad x_{2n} = 2x_n^2 - 1.$$

When $d = 3$, we also have $x_{n+1} - y_{n+1} = x_n + y_n$. The solutions of $x^2 - 7y^2 = 1$ satisfy

$$x_{2n-1} = (y_n - y_{n-1})^2 - 1, \quad x_{2n} = 2x_n^2 - 1,$$

while those of $x^2 - 8y^2 = 1$ satisfy

$$x_{2n-1} = [2(y_n - y_{n-1})]^2 - 1, \quad x_{2n} = [4y_n]^2 - 1.$$

Chapter 3

1.1(a). $-78 + 13\sqrt{3}$.

1.1(b). $-143, -39, 5577$.

1.1(c). $-(2/143) + (7/143)\sqrt{3}$.

1.1(d). $(2 + \sqrt{3})^{-1} = 2 - \sqrt{3}$.

1.7. Some are $(x, y) = (7, 5), (41, 29)$.

1.8. Some are $(x, y) = (13, 9), (75, 53)$.

1.9(c). $(r, s) = (1, 1)$.

1.9(e). Observe that

$$\left(r\sqrt{w+\frac{1}{2}}+s\sqrt{w-\frac{1}{2}}\right)\left(r\sqrt{w+\frac{1}{2}}-s\sqrt{w-\frac{1}{2}}\right)$$
$$=\left(w+\frac{1}{2}\right)r^2-\left(w-\frac{1}{2}\right)s^2,$$

so that we can obtain solutions to (b) by looking at surd powers. Using (c) and the fact that

$$\left(r\sqrt{w+\frac{1}{2}}+s\sqrt{w-\frac{1}{2}}\right)\left(\sqrt{w+\frac{1}{2}}+\sqrt{w-\frac{1}{2}}\right)^2$$
$$=\left(r\sqrt{w+\frac{1}{2}}+s\sqrt{w-\frac{1}{2}}\right)\left(2w+2\sqrt{w+\frac{1}{2}}\sqrt{w-\frac{1}{2}}\right)$$
$$=[2wr+(2w-1)s]\sqrt{w+\frac{1}{2}}+[(2w+1)+2ws]\sqrt{w-\frac{1}{2}}$$

we obtain the answer.

(i) That y_n^2+v is a multiple of $v+1$ can be established by induction from (h). Use (g) to establish that both equations of the system hold.

(j) $(x, y, u, v) = (v, v, v, v)$, $(2v^2-v, 2v^2+v, 4v^3+v, v)$, $(4v^3-2v^2-v, 4v^3+2v^2-v, 16v^5-4v^3+v, v)$, ... When $v=1$, then x^2 must be equal to 1, and $y=2z+1$ is odd. Then $u=2z^2+2z+1$, and we obtain the solution

$$(x, y, u, v) = (\pm 1, 2z+1, 2z^2+2z+1, 1).$$

Note that $x=1+2+\cdots+(2v-1)=v(2v-1)$ and $y=1+2+\cdots+2v=v(2v+1)$ corresponds to the second of our series of solutions.

1.10(a). Possible values for (m, r) are $(1, 24)$, $(2, 40)$, $(3, 56)$, $(4, 72)$. If $r=8+16m$, then

$$mr+1=(4m+1)^2, \qquad (m+1)r+1=(4m+3)^2.$$

1.10(b). $(m+1)x^2-my^2=1$ is equivalent to $(m+1)(x^2-1)=m(y^2-1)$. Now use the fact that m and $m+1$ are coprime.

1.10(c). An obvious solution is $(x_1, y_1) = (1, 1)$. Other solutions are given by

$$\sqrt{m+1}x_{n+1}+\sqrt{m}y_{n+1}=\left(\sqrt{m+1}x_n+\sqrt{m}y_n\right)\left(\sqrt{m+1}+\sqrt{m}\right)^2.$$

Thus

$$x_{n+1}=(2m+1)x_n+2my_n,$$
$$y_{n+1}=2(m+1)x_n+(2m+1)y_n.$$

1.10(d).

$$\begin{aligned} r_{n+1} &= \frac{x_{n+1}^2 - 1}{m} \\ &= r_n + 4[(m+1)x_n^2 + my_n^2 + (2m+1)x_n y_n] \\ &= r_n + 4(x_n + y_n)[(m+1)x_n + my_n] \\ &= (8m^2 + 8m + 1)r_n + 4(2m+1)(x_n y_n + 1). \end{aligned}$$

2.1(a). $(x, y) = (c^2 + d, 2c)$.

2.1(b).

$$(x, y) = \left(\frac{c^2 + d}{c^2 - d}, \frac{2c}{c^2 - d}\right).$$

2.2(a). Since $(x-3)(x+3) = 13y^2$, x should be 3 more or 3 less than a multiple of 13. Thus we try $x = 3, 10, 16, 23, 29$. We have that $(29-3)(29+3) = 13 \times 2^6$, so $(x, y) = (29, 8)$ works. A rational solution of $x^2 - 13y^2 = 1$ is $(x, y) = (\frac{29}{3}, \frac{8}{3})$.

2.4(a). Consider $(7 + \sqrt{61})/(8 - \sqrt{61})$.

2.6(a). $x^2 - 19y^2 = 6$ is satisfied by $(x, y) = (5, 1)$, so $x^2 - 19y^2 = 36$ is satisfied by $(x, y) = (44, 10)$. The desired result follows from this.

2.6(b). $x^2 - 19y^2 = -3$ is satisfied by $(x, y) = (4, 1)$, and the result follows from this.

2.6(c). Observe that the coefficients of $(35 + 8\sqrt{19})(22 + 5\sqrt{19})$ are divisible by 9, so that

$$\frac{35 + 8\sqrt{19}}{22 - 5\sqrt{19}} = 170 + 39\sqrt{19}.$$

Thus, $(x, y) = (170, 39)$ satisfies $x^2 - 19y^2 = 1$. Indeed, $170^2 - 1 = 171 \times 169 = 19 \times 3^2 \times 13^2$.

3.4. $\cos 2\theta = 2\cos^2\theta - 1$; $\cos 3\theta = 4\cos^3\theta - 3\cos\theta$, $\cos 4\theta = 8\cos^4\theta - 8\cos^2\theta + 1$.

3.5. $\sin 2\theta = 2\sin\theta\cos\theta$, $\sin 3\theta = \sin\theta[4\cos^2\theta - 1] = 3\sin\theta - 4\sin^3\theta$, $\sin 4\theta = 4\sin\theta[2\cos^3\theta - \cos\theta]$

4.2(a). Let $t = \cos\theta$ with $0 \le \theta \le \pi$. Note that $\cos n\theta + \cos(n-2)\theta = 2\cos\theta\cos(n-1)\theta$.

4.2(d). $T_n(t) = 0$ if and only if $n\theta$ is an odd multiple of $\pi/2$. This happens for n values of θ satisfying $0 \le \theta \le \pi$, namely $\theta = \pi/2n, 3\pi/2n, \ldots, (2n-1)\pi/2n$. All of these yield different values of $t = \cos\theta$.

4.3(a). Let $t = \cos\theta$ with $0 \leq \theta \leq \pi$. Then $\sin n\theta + \sin(n-2)\theta = 2\cos\theta\sin(n-1)\theta$ so that

$$\begin{aligned} U_n(t) &= (\sin\theta)^{-1}\sin n\theta = (\sin\theta)^{-1}[2\cos\theta\sin(n-1)\theta - \sin(n-2)\theta] \\ &= 2\cos\theta[(\sin\theta)^{-1}\sin(n-1)\theta] - (\sin\theta)^{-1}\sin(n-2)\theta \\ &= 2tU_{n-1}(t) - U_{n-2}(t). \end{aligned}$$

Here is a table of the first few Chebyshev polynomials:

n	$T_n(t)$	$U_n(t)$
0	1	0
1	t	1
2	$2t^2 - 1$	$2t$
3	$4t^3 - 3t$	$4t^2 - 1$
4	$8t^4 - 8t^2 + 1$	$8t^3 - 4t$
5	$16t^5 - 20t^3 + 5t$	$16t^4 - 12t^2 + 1$
6	$32t^6 - 48t^4 + 18t^2 - 1$	$32t^5 - 32t^3 + 6t$

4.5. A straightforward induction argument using the recursions in Exercises 4.2 and 4.3 yields that $2T_{n-1}(t) = U_n(t) - U_{n-2}(t)$ for $n \geq 1$. The exercise can be solved by induction, the induction step being

$$\begin{aligned} T_n'(t) &= 2T_{n-1}(t) + 2tT_{n-1}'(t) - T_{n-2}'(t) \\ &= 2T_{n-1}(t) + 2(n-1)tU_{n-1}(t) - (n-2)U_{n-2}(t) \\ &= 2T_{n-1}(t) + [-2tU_{n-1}(t) + 2U_{n-2}(t)] + n[2tU_{n-1}(t) - U_{n-2}(t)] \\ &= nU_n(t) + [2T_{n-1}(t) - U_n(t) + U_{n-2}(t)] = nU_n(t). \end{aligned}$$

4.6(a). Use

$$\cos 2n\theta = 1 - 2\sin^2 n\theta = 1 - 2\sin^2\theta\left(\frac{\sin n\theta}{\sin\theta}\right)^2.$$

4.6(b). Use $\sin 2n\theta = 2\sin n\theta\cos n\theta$.

4.7. Since

$$\begin{aligned} &(1+\cos\theta)\cos(2n+1)\theta \\ &\quad = (1+\cos\theta) + (1+\cos\theta)(\cos(2n+1)\theta - 1) \\ &\quad = (1+\cos\theta) + \left(2\cos^2\frac{\theta}{2}\right)\left(-2\sin^2\left(\frac{2n+1}{2}\right)\theta\right) \\ &\quad = (1+\cos\theta) - \left(2\cos\frac{\theta}{2}\sin\frac{2n+1}{2}\theta)\right)^2 \\ &\quad = (1+\cos\theta) - [\sin(n+1)\theta + \sin n\theta]^2, \end{aligned}$$

we have

$$\cos(2n+1)\theta = 1 - \frac{\sin^2\theta}{1+\cos\theta}\left[\frac{\sin(n+1)\theta}{\sin\theta} + \frac{\sin n\theta}{\sin\theta}\right]^2.$$

4.8(a). Let $t = \cos\theta$, so $T_n(t) = \cos n\theta$. Then

$$T_m(T_n(\theta)) = \cos m(\arccos(\cos n\theta)) = \cos m(n\theta)) = T_{mn}(t).$$

4.8(b). $U_m(T_n(t)) = \dfrac{\sin m(\arccos(\cos n\theta))}{\sin(\arccos(\cos n\theta))} = \dfrac{\sin mn\theta}{\sin n\theta}$, so that

$$U_m(T_n(t))U_n(t) = \frac{\sin mn\theta}{\sin n\theta}\cdot\frac{\sin n\theta}{\sin\theta} = \frac{\sin mn\theta}{\sin\theta} = U_{mn}(t).$$

4.9(a). $(z+w)^2 = z^2+2wz+w^2 = (2z^2-1)+2zw$; $(z+w)^3 = z^3+3z^2w+3zw^2+w^3 = z^3+3z(z^2-1)+(3z^2+z^2-1)w = (4z^3-3z)+(4z^2-1)w$.

4.9(b). In general, $(z+w)^n = T_n(z)+U_n(z)w$. This can be established by induction, using the recursion in Exercise 4.4(a).

4.9–4.10.

$$\begin{aligned}\cos n\theta + i\sin n\theta &= \sum_{i=0}^{\lfloor n/2\rfloor}(-1)^i\binom{n}{2i}\cos^{n-2i}\theta\sin^{2i}\theta \\ &\quad + i\sin\theta\sum_{j=0}^{\lfloor n/2\rfloor}(-1)^j\binom{n}{2j+1}\cos^{n-2j-1}\theta\sin^{2j}\theta \\ &= \sum_{i=0}^{\lfloor n/2\rfloor}(-1)^i\binom{n}{2i}\cos^{n-2i}\theta(1-\cos^2\theta)^i \\ &\quad + i\sin\theta\sum_{j=0}^{\lfloor n/2\rfloor}(-1)^j\binom{n}{2j+1}\cos^{n-2j-1}\theta(1-\cos^2\theta)^j\end{aligned}$$

yields

$$T_n(t) = \sum_{i=0}^{\lfloor n/2\rfloor}\binom{n}{2i}(-1)^i t^{n-2i}(1-t^2)^i = \sum_{i=0}^{\lfloor n/2\rfloor}\binom{n}{2i}t^{n-2i}(t^2-1)^i$$

and

$$U_n(t) = \sum_{j=0}^{\lfloor n/2\rfloor}(-1)^j\binom{n}{2j+1}t^{n-2j-1}(1-t^2)^j = \sum_{j=0}^{\lfloor n/2\rfloor}\binom{n}{2j+1}t^{n-2j-i}(t^2-1)^j.$$

It follows that

$$\begin{aligned}T_n(z)+U_n(z)w &= \sum_{i=0}^{\lfloor n/2\rfloor}\binom{n}{2i}z^{n-2i}w^{2i} + \sum_{j=0}^{\lfloor n/2\rfloor}\binom{n}{2j+1}z^{n-2j-1}w^{2j+1} \\ &= \sum_{k=0}^{n}\binom{n}{k}z^{n-k}w^k = (z+w)^n.\end{aligned}$$

4.11. This is a direct consequence of Exercise 4.9 with $z = u$ and $w = v\sqrt{d}$.

4.12(b). There are at most m^2 possible pairs (a_n, a_{n+1}) modulo m, so that by the pigeonhole principle, any collection of $m^2 + 1$ pairs must contain two, (a_r, a_{r+1}) and (a_s, a_{s+1}), that are congruent modulo m: $a_r \equiv a_s$ and $a_{r+1} \equiv a_{s+1}$.

4.12(c). By induction, it can be shown from the recursion $a_{n+1} + a_{n-1} = 2ca_n$ that $a_{r+k} \equiv a_{s+k}$ for each integer k. In particular, it holds for $k = -r$.

4.13. From Exercises 4.12 and 4.3 we can select an integer n such that $U_n(u) \equiv U_0(u) = 0$; $U_{n+1} \equiv U_1(u) = 1$. Then $T_n(u) \equiv U_{n+1}(u) - uU_n = 1$.

4.14. Let $m^2n+1 = y^2$ and $(m^2-1)n+1 = z^2$. Then $m^2z^2-(m^2-1)y^2 = 1$ and $n = y^2-z^2$. Setting $x = mz$, we see that we seek solutions to $x^2-(m^2-1)y^2 = 1$ with x a multiple of m. From Exercise 4.1 we determine solutions $(x_r, y_r) = (T_r(m), U_r(m))$ for $r = 0, 1, 2, 3, \ldots, m$, and from Exercise 4.2(e), we note that x_r is divisible by m if and only if r is odd; these yield values $z_r = x_r/m$ and $n_r = y_r^2 - z_r^2$. Observe the initial cases, $(r, n_r; z_r, y_r) = (1, 0; 1, 1), (3, 8(2m^2 - 1); 4m^2 - 3, 4m^2 - 1)$.

We show that n_r is divisible by $8(2m^2 - 1) = n_3$ for all odd values of r. This is clear for n_1 and n_3. Since

$$x_{r+2} + y_{r+2}\sqrt{m^2 - 1} = (m + \sqrt{m^2 - 1})^2(x_r + y_r\sqrt{m^2 - 1}),$$

we find that

$$\begin{aligned} z_{r+2} &= (2m^2 - 1)z_r + 2(m^2 - 1)y_r, \\ y_{r+2} &= 2m^2z_r + (2m^2 - 1)y_r. \end{aligned}$$

It is straightforward to prove by induction that z_r and y_r are both odd for each odd value of r, so that $y_r + z_r$ is a multiple of 2. The desired result not follows from an induction argument using

$$\begin{aligned} n_{r+2} = y_{r+2}^2 - z_{r+2}^2 &= (y_{r+2} - z_{r+2})(y_{r+2} + z_{r+2}) \\ &= (y_r + z_r)[(4m^2 - 1)z_r + (4m^2 - 3)y_r] \\ &= (y_r + z_r)[(4m^2 - 2)(y_r + z_r) - (y_r - z_r)] \\ &= 2(2m^2 - 1)(y_r + z_r)^2 - (y_r^2 - z_r^2) \\ &= n_3\left[\frac{1}{2}(y_r + z_r)\right]^2 - n_r. \end{aligned}$$

The solution of the problem follows from the fact that $(x, y) = (m, 1)$ is a fundamental solution of $x^2 - (m^2 - 1)y^2 = 1$, so that all values of n are given by the n_r. (See Section 4.2.)

5.1. Suppose that $(x, y) = (r, s)$ satisfies $x^2 - dy^2 = k$. Then $(x, y) = (r, ms)$ satisfies $x^2 - ey^2 = k$. On the other hand, let $(x, y) = (u, v)$ be a solution of $x^2 - ey^2 = k$. By Exercise 4.12, we can select n such that $U_n(u)$ is divisible by m. Then $(x, y) = (T_n(u), vU_n(u)/m)$ is a solution of $x^2 - dy^2 = k$.

5.1(b). If (u, v) is the smallest solution of $x^2 - ey^2 = 1$, then from (a) and Exercise 4.10, every solution of $x^2 - dy^2 = 1$ is given by $(x, y) = (T_n(u), vU_n(u))$, where $U_n(u)$ is divisible by m.

5.2. The smallest solutions of $x^2 - 2y^2 = 1$ are (3, 2), (17, 12), (99, 70), (577, 408), (3363, 2378), (19601, 13860).

$d = 8$ (3, 1); $d = 18$ (17, 4); $d = 32$ (17, 3); $d = 50$ (99, 14); $d = 72$ (17, 2); $d = 98$ (99, 10); $d = 128$ (577, 51); $d = 162$ (19601, 1540); $d = 200$ (99, 7).

5.3. The smallest solutions of $x^2 - 3y^2 = 1$ are (2, 1), (7, 4), (26, 15), (97, 56), (362, 209), (1351, 780).

$d = 12$ (7, 2); $d = 27$ (26, 5); $d = 48$ (7, 1); $d = 75$ (26, 3); $d = 108$ (1351, 130).

Comments on the Explorations

Exploration 3.1. To make the computations more palatable, let $q = 2m + 1$. Then

$$\begin{pmatrix} x_{n+1} \\ y_{n+1} \end{pmatrix} = \begin{pmatrix} q & q-1 \\ q+1 & q \end{pmatrix} \begin{pmatrix} x_n \\ y_n \end{pmatrix}.$$

Using a method similar to that of Exercise 1.2.3, we find that

$$x_{n+1} = 2qx_n - x_{n-1},$$
$$y_{n+1} = 2qy_n - y_{n-1},$$

for $n \geq 2$. The sequence $(x_n, y_n; r_n)$ contains the terms

$$(1, 1; 0), (2q-1, 2q+1; 8q), (4q^2 - 2q - 1, 4q^2 + 2q - 1; 8q(4q^2 - 1)),$$
$$(8q^3 - 4q^2 - 4q + 1, 8q^3 + 4q^2 - 4q - 1; 16q(2q^2 - 1)(4q^2 - 1)),$$
$$(16q^4 - 8q^3 - 12q^2 + 4q + 1, 16q^4 + 8q^3 - 12q^2 - 4q + 1;$$
$$16q(2q^2 - 1)[2(4q^2 - 1)(2q^2 - 1) - 1],$$
$$(32q^5 - 16q^4 - 32q^3 + 12q^2 + 6q - 1, 32q^5 + 16q^4 - 32q^3 - 12q^2 + 6q + 1;$$
$$8q(4q^2 - 1)(4q^2 - 3)[2(2q^2 - 1)(4q^2 - 1) - 1]).$$

Exploration 3.2. Here are a few values: $(d, k; x, y) = (2, 1.4; 99, 70)$, $(3, 1.75; -97, -56)$, $(5, 2.25; -161, -72)$, $(6, 2.5; -49, -20)$, $(6, 2.4; 49, 20)$, $(12, 3.5; -97, -28)$. If we take $d = 61$ and $k = 7.8$, we find the solution $(x, y) = (761.5, 97.5)$; can this be parlayed to an integer solution?

Chapter 4

1.4. There are exactly k^2 pairs (u, v) for which $0 \leq u, v \leq k - 1$. For each solution (x_i, y_i), there is exactly one such pair (u_i, v_i) for which $x_i \equiv u_i$ and

$y_i \equiv v_i \pmod{k}$. Since there are more solutions than pairs, by the pigeonhole principle there must be some pair to which two solutions correspond. The result follows.

1.6(a). Apply the pigeonhole principle. Since there are more numbers than intervals, two of the numbers must fall in the same interval.

1.6(d).

$$|u^2 - dv^2| = |u - v\sqrt{d}||u + v\sqrt{d}| < \frac{1}{v}\left[\frac{1}{v} + 2v\sqrt{d}\right] = 2\sqrt{d} + \frac{1}{v^2} < 2\sqrt{d} + 1.$$

1.7. Suppose finitely many pairs $(u_1, v_1), (u_2, v_2), \ldots, (u_m, v_m)$ with the desired property have been found. Let N be a positive integer with $1/N$ less than $|u_i - v_i\sqrt{d}|$ for each i. Then, as in Exercise 1.6, select u_{m+1} and v_{m+1} such that

$$\left|u_{m+1} - v_{m+1}\sqrt{d}\right| < \frac{1}{N} \le \frac{1}{v_{m+1}}.$$

Since $|u_{m+1} - v_{m+1}\sqrt{d}| < |u_i - v_i\sqrt{d}|$ for $1 \le i \le m$, we have increased the number of pairs. The result follows.

1.8. By Exercise 1.7, there are infinitely many pairs (u, v) for which $|u - v\sqrt{d}| < 1/v$, and for each pair $|u^2 - dv^2| \le 2\sqrt{d} + 1$. There are finitely many integers between $-(2\sqrt{d}+1)$ and $2\sqrt{d}+1$, so by an extension of the pigeonhole principle, at least one of these must be assumed infinitely often by $x^2 - dy^2$.

1.10(a). $u^2z^2 - v^2w^2 = (k + dv^2)z^2 - v^2(k + dz^2)$.

2.3(a). $u^2 - dv^2 = x_1^2 - dy_1^2 \Rightarrow (u^2 - x_1^2) = d(v^2 - y_1^2)$. Since $u \ge x_1$ by the definition of x_1, we must have $v \ge y_1$, so that $u + v\sqrt{d} \ge x_1 + y_1\sqrt{d}$.

2.6. The $(n+1)$th approximant is x_{2^n}/y_{2^n}, as can be verified by induction.

3.1. Any nonsquare d can be written in the form em^2 where e is a squarefree number, namely, the product of primes that divide d to an odd power. Any number of the form $r + s\sqrt{d}$ can be written as $r + sm\sqrt{e}$.

3.2(a). Suppose $r = m/n$ in lowest terms for integers m and n. Then $m^2 + bmn + cn^2 = 0$, so that any prime divisor of n must divide m^2 and hence m. Since m and n are relatively prime, n must be equal to 1.

3.2(b). A similar argument holds for polynomials in general.

3.3. $u + v\sqrt{d}$ is a root of the monic quadratic $t^2 - 2ut + (u^2 - v^2d)$; the other root is $u - v\sqrt{d}$.

3.4(a). $(u^2 + dv^2 + 2uv\sqrt{d}) + bu + bv\sqrt{d} + c = 0$ so that $v(2u + b) = 0$ and $u^2 + dv^2 + bu + c = 0$. Hence $2u = -b$, so that $4dv^2 = -4(u^2 + bu + c) = -4(u^2 - 2u^2 + c) = 4u^2 - 4c = b^2 - 4c$.

3.4(b). Suppose $2v = m/n$ in lowest terms. Then, since $4dv^2 = dm^2/n^2$ is an integer, n^2 must divide d. But d is squarefree; hence $n = 1$.

3.4(c). If b is odd, then $(2v)^2d$ must be odd, so that both $2v$ and d are odd. Since $1 \equiv b^2 \equiv (2v)^2d \pmod{4}$ and $(2v)^2 \equiv 1 \pmod{4}$, we must have $d \equiv 1 \pmod{4}$.

3.5. Suppose that $u + v\sqrt{d}$ is a quadratic integer. Then $2u$ is even or odd. If $2u$ is odd, so is $2v$, and $d \equiv 1 \pmod{4}$, by Exercise 3.4(c). If $2u$ is even, so are b, and $4dv^2 = b^2 - 4c = 4(u^2 - c)$. Since d, being square-free, is not divisible by 4, $2v$ must be even. Hence $2u$ and $2v$ must be even. We conclude that $2u$ and $2v$ have the same parity, and the necessary condition follows. The converse is straightforward.

3.7. Suppose $u^2 - dv^2 = 1$, where $u + v\sqrt{d}$ is a quadratic integer. Then $(x, y) = (2u, 2v)$ are solutions of $x^2 - dy^2 = 4$ with integers of the same parity. On the other hand, suppose that $r^2 - ds^2 = 4$ with r and s of the same parity. Then $(x, y) = (r/2, s/2)$ is a solution of $x^2 - dy^2 = 1$. Suppose r and s are even; then $r/2$ and $s/2$ are ordinary integers. Suppose r and s are odd. Then $\frac{1}{2}(r + s\sqrt{d})$ is a root of the monic quadratic polynomial

$$t^2 - rt + \frac{r^2 - ds^2}{4}$$

and so is a quadratic integer.

3.9. Solutions of $x^2 - dy^2 = 4$ with x, y odd:

d	(x, y)
5	(3, 1), (7, 3)
13	(11, 3), (119, 33)
21	(5, 1), (23, 5)
29	—
37	—
45	(7, 1), (47, 7)
53	(51, 7), (2599, 357)
61	(1523, 195), (2319527, 296985)
69	(25, 3), (623, 75)
77	(9, 1), (79, 9)

There are no solutions when $d = 29, 37, 83, 91, 99$.

3.11(b). Suppose $x^2 - dy^2 = -4$ with x and y odd and positive. Then $(x/2) + (y/2)\sqrt{d}$ is a unit, and so must be a power of the minimum positive unit $(r/2) + (s/2)\sqrt{d}$, whose norm must be -1. $(m/2) + \frac{1}{2}\sqrt{d}$ must be a power of the same unit. Since $s/2 > \frac{1}{2}$, this cannot occur.

3.13. $u = p(p^2 + 3dq^2)/8$ and $v = q(3p^2 + dq^2)/8$. Since p and q are both odd and $d \equiv 5 \pmod{8}$, $p^2 + 3q^2d \equiv 1 + 15 \equiv 0$ and $3p^2 + dq^2 \equiv 3 + 5 \equiv 0 \pmod{8}$. Thus u and v are both integers.

4.2(a). Observe that s_0 is an integer, and that $p(s_0, z)$ is a monic quadratic polynomial with integer coefficients and at least one integer root. $ds_0^2 + k = ds^2 + k = r^2$, so that the result holds for $n = 0$. Suppose $r_n^2 = ds_n^2 + k$ for some nonnegative integer n. Then

$$\begin{aligned} 0 &= s_{n+1}^2 - 2us_{n+1}s_n + s_n^2 - v^2k \\ &= s_{n+1}^2 - 2us_{n+1}s_n + u^2s_n^2 - v^2(ds_n^2 + k) \\ &= (s_{n+1} - s_n u)^2 - v^2r_n^2, \end{aligned}$$

so that $s_{n+1} = s_n u \pm r_n v$. Let $r_{n+1} = r_n u \pm ds_n v$. Then

$$\begin{aligned} r_{n+1}^2 - ds_{n+1}^2 &= r_n^2u^2 + d^2s_n^2v_n^2 - ds_n^2u^2 - dr_n^2v^2 \\ &= (r_n^2 - ds_n^2)(u^2 - dv^2) = k, \end{aligned}$$

from which we find that $ds_{n+1}^2 + k$ is a square. The result follows for nonnegative n by induction. Working "backwards" will give the result for negative n as well.

4.2(b). s_{n+1} and s_{n-1} are roots of a quadratic polynomial the sum of whose roots is $2us_n$ and product is $s_n^2 - v^2k$, as read from the coefficients.

4.3(a). Suppose $s_n > 0$. Then $s_{n+1} + s_{n-1}$ and $s_{n+1}s_{n-1}$ are both positive, as will be s_{n+1} and s_{n-1}.

4.3(b). Note that $s_{n+1}s_{n-1} - s_n^2 > 0$. The latter part follows by induction from $s_1/s_0 \geq 1$ and $s_0/s_{-1} \leq 1$.

4.4. From Exercise 4.3(c) we have that

$$s_0 \leq s_1 = us_0 + v\sqrt{ds_0^2 + k} \quad \text{and} \quad s_0 \leq s_{-1} = us_0 - v\sqrt{ds_0^2 + k},$$

from which

$$-(u-1)s_0 \leq v\sqrt{ds_0^2 + k} \leq (u-1)s_0,$$

whence

$$|v\sqrt{ds_0^2 + k}| \leq (u-1)s_0.$$

Squaring yields the inequality

$$\begin{aligned} &dv^2s_0^2 + v^2k \leq (u^2 - 2u + 1)s_0^2 \\ &\Longrightarrow v^2k \leq (u^2 - dv^2 - 2u + 1)s_0^2 = -2(u-1)s_0^2 \\ &\Longrightarrow 2(u-1)s_0^2 \leq -v^2k = v^2|k| \\ &\Longrightarrow s_0 \leq v\sqrt{\frac{|k|}{2(u-1)}}. \end{aligned}$$

4.5(a). $p(s_n, z) = 0$ has a double root if and only if

$$\begin{aligned} 0 &= 4u^2s_n^2 - 4(s_n^2 - v^2k) = 4s_n^2(u^2 - 1) + 4v^2k \\ &= 4s_n^2dv^2 + 4v^2k = 4v^2(ds_n^2 + k), \end{aligned}$$

which is equivalent to $k = -ds_n^2$.

4.6(a). $p(y, y) = 2y^2(1-u) - v^2k \leq 0$, since $u \geq 1$.

4.6(b). From (a), s_0 must lie between two solutions s_1 and s_{-1} of $p(s_0, z) = 0$. Note that $s_{n+1} - s_n = (2u-1)(s_n - s_{n-1}) + 2(u-1)s_{n-1}$. If $n \geq 1$, it can be shown by induction that $s_{n+1} \geq s_n$. Since s_0 is the smallest nonnegative term and lies between s_1 and s_{-1}, we have that $s_{-1} < 0$. Since $s_n - s_{n-1} = -2(u-1)s_n + (s_{n+1} - s_n)$, it can be shown by induction that $s_n \geq s_{n-1}$ for all $n \leq -1$.

4.6(c). Since $s_1 s_{-1} < 0$, $s_0^2 - v^2 k < 0$. If $x^2 - ds_0^2 = k$, then $x^2 = k + ds_0^2 < k(1 + dv^2) = ku^2$.

4.6(d). When $n \neq 0, -1$, $s_n^2 - v^2 k = s_{n-1}s_{n+1} \geq 0$.

4.7(a).

$$d^2v^2s^2 = (dv^2)(ds^2) = (u^2-1)(r^2-k) < u^2r^2.$$

Since all variables are positive, this implies that $dvs < ur$.

4.7(b). Since r is minimal, by (a) we must have that $ur - dvs \geq r$, whence $r(u-1) \geq dvs$. Squaring this yields $r^2(u-1)^2 \geq d^2v^2s^2 = (u^2-1)(r^2-k)$, so that $r^2(u-1) \geq (u+1)(r^2-k)$ (since $u \neq 1$).

4.7(d).

$$\frac{s^2(u^2-1)}{v^2} = ds^2 = r^2 - k \leq \frac{k(u+1)}{2} - k = \frac{k(u-1)}{2},$$

whence

$$s^2 \leq \frac{kv^2}{2(u+1)}.$$

4.7(e). Since $x^2 = dy^2 + k$, the smallest value of $|x|$ among solutions occurs along with the smallest value of $|y|$. Thus, y is equal to s_0 or s_{-1}, depending on whether it is positive or negative.

4.8. In this case, $(u, v) = (3, 2)$ and $p(y, z) = y^2 - 6yz + z^2 - 28$.

4.9. In this case, $p(y, z) = y^2 - 6yz + z^2 + 28$.

Comments on the Explorations

Exploration 4.1. Most general texts in elementary number theory, such as those by LeVeque; Mollin; Niven, Zuckerman and Montgomery; Rosen, cover the Legendre symbol. For odd primes $(-1/p) = (-1)^{(p-1)/2}$ and $(2/p) = (-1)^{(p^2-1)/8}$. Thus, $x^2 \equiv -1 \pmod{p}$ is solvable if and only if $p \equiv 1 \pmod 4$, and $x^2 \equiv 2 \pmod p$ is solvable if and only if $p \equiv \pm 1 \pmod 8$. A deep and useful rule in the computation of Legendre symbols is the law of quadratic reciprocity:

$$\left(\frac{p}{q}\right)\left(\frac{q}{p}\right) = (-1)^{\frac{p-1}{2}\cdot\frac{q-1}{2}}$$

for odd primes p and q. For example,

$$\left(\frac{19}{37}\right)\left(\frac{37}{19}\right) = (-1)^{9\times 18} = 1,$$

with the result that $\left(\frac{19}{37}\right) = \left(\frac{37}{19}\right) = \left(\frac{-1}{19}\right) = -1$, so that $x^2 \equiv 19 \pmod{37}$ and $x^2 - 37y^2 = 19$ are not solvable for integers x and y.

Exploration 4.2. The solutions are $(x, y) = (2, 0), (1, 1), (3, 1), (4, 2), (7, 3), (11, 5), \ldots$, where the values of y seem to run through the Fibonacci numbers and the values of x through the Lucas numbers L_n. The sequences are defined by

$$L_0 = 2, \quad L_1 = 1, \quad L_{n+1} = L_n + L_{n-1} \quad \text{(Lucas sequence);}$$
$$F_0 = 0, \quad F_1 = 1, \quad F_{n+1} = F_n + F_{n-1} \quad \text{(Fibonacci sequence).}$$

Note that $L_n = F_{n-1} + F_{n+1}$ for $n \geq 1$. One conjectures that

$$(F_{n-1} + F_{n+1})^2 - 5F_n^2 = (-1)^n$$

and

$$L_n L_{n+1} - 5F_n F_{n+1} = 2(-1)^n$$

for $n \geq 0$. Indeed, the left side of the equation is

$$(2F_{n-1}+F_n)^2-5F_n^2 = 4[F_{n-1}^2+F_{n-1}F_n - F_n^2] = 4[F_{n-1}F_{n+1} - F_n^2] = 4(-1)^n.$$

Exploration 4.3. By examining the continued fraction development of $\sqrt{d}$ to be discussed in Chapter 5, we find that some values of d for which there are no solutions of the type desired are numbers of the form $(4k + 2)^2 + 1$, while there are others such as 141, 189, 269, 333, 349, 373, 381, 389, 405.

Exploration 4.5. The theory of quadratic forms is a long-established and well-developed area of mathematics. Gauss was the first to embark on a systematic study in *Disquisitiones arithmeticae*, written in 1801. An English translation is now available. Harold Davenport's *The Higher Arithmetic* has a chapter (VI) on the theory of quadratic forms. Duncan A. Buelle, in his *Binary Quadratic Forms: Classical Theory and Modern Computation*, provides a summary of the basic theory that is suitable for undergraduates with some background in modern algebra; he gives a good treatment of computational issues. Quadratic forms $ax^2+bxy+cy^2$ with discriminant $\Delta = b^2 - 4ac$ can be put into finitely many equivalence classes determined by linear transformations of the variables, and these equivalence classes themselves can be given an algebraic structure.

Exploration 4.6. Let $(x, y) = (u, v)$ be a particular solution of $x^2 - dy^2 = 1$. As we have seen in Sections 3.3 and 3.4, $x_n^2 - dy_n^2 = 1$ holds when $(x_n, y_n) = (T_n(u), U_n(u)v)$. Let $T_n(u)U_n(u) = S_n(u)$, so that $z_n = S_n(u)v$; note that $T_n(u)$ is divisible by u when n is odd, and $U_n(u)$ is divisible by u when n is even. Empirical evidence suggests that

$$S_{n+1}(u) = [(2u)^2 - 2]S_n(u) - S_{n-1}(u) \quad \text{for } n \geq 1$$

and that $2S_n = U_{2n}$. The latter is a consequence of Exercise 3.4.6(b), since $S_n = T_nU_n$. To obtain the recursion, note that

$$\begin{aligned} T_{n+1}U_{n+1} &= (2uT_n - T_{n-1})(2uU_n - U_{n-1}) \\ &= (4u^2 - 2)T_nU_n - T_{n-1}U_{n-1} \\ &\quad + 2[(T_nU_n + T_{n-1}U_{n-1}) - u(T_nU_{n-1} + T_{n-1}U_n)] \end{aligned}$$

and that

$$\begin{aligned} &2[(T_nU_n + T_{n-1}U_{n-1}) - u(T_nU_{n-1} + T_{n-1}U_n)] \\ &\quad = T_n(U_n - 2uU_{n-1}) + U_n(T_n - 2uT_{n-1}) + 2T_{n-1}U_{n-1} \\ &\quad = 2T_{n-1}U_{n-1} - T_nU_{n-2} - U_nT_{n-2} \\ &\quad = 2[(T_{n-1}U_{n-1} + T_{n-2}U_{n-2}) - u(T_{n-1}U_{n-2} + U_{n-1}T_{n-2})], \end{aligned}$$

so that by induction it can be shown that $2[\cdots]$ vanishes.

Let $R_n = T_{2n}$. Then $R_n^2 - 4d^2S_n^2v^2 = T_{2n}^2 - d^2(U_{2n}v)^2 = 1$.

A direct argument that does not use Chebyshev polynomials follows:

For an arbitrary value of d, the analogue of the result in Exercise 1.2.3 is

$$\begin{aligned} x_{n+1} &= x_1x_n + dy_1y_n, \\ y_{n+1} &= y_1x_n + x_1y_n, \end{aligned}$$

with corresponding matrix $\begin{pmatrix} x_1 & dy_1 \\ y_1 & x_1 \end{pmatrix}$. One finds that, for $n \geq 1$,

$$\begin{aligned} x_{n+1} &= 2x_1x_n - x_{n-1}, \\ y_{n+1} &= 2x_1y_n - y_{n-1}. \end{aligned}$$

Then

$$x_{n+1}y_{n+1} = 4x_1^2x_ny_n - 2x_1(x_ny_{n-1} + x_{n-1}y_n) + x_{n-1}y_{n-1}.$$

Now,

$$\begin{aligned} x_1x_ny_{n-1} + x_1x_{n-1}y_n &= x_n(y_n - y_1x_{n-1}) + x_{n-1}x_1y_n \\ &= x_ny_n - x_{n-1}(y_1x_n - x_1y_n) \end{aligned}$$

so

$$\begin{aligned} x_{n+1}y_{n+1} &= (4x_1^2 - 2)x_ny_n + x_{n-1}(2y_1x_n - 2x_1y_n + y_{n-1}) \\ &= (4x_1^2 - 2)x_ny_n + x_{n-1}(2y_1x_n - y_{n+1}) \\ &= (4x_1^2 - 2)x_ny_n + x_{n-1}(y_{n+1} - 2x_1y_n) \\ &= (4x_1^2 - 2)x_ny_n - x_{n-1}y_{n-1}. \end{aligned}$$

We now look for a number e for which $w_n^2 - ez_n^2 = 1$ for each value of n. From $(x_2, y_2) = (x_1^2 + dy_1^2, 2x_1y_1) = (2x_1^2 - 1, 2x_1y_1)$, we have that $z_2 = 2x_1y_1(2x_1^2 - 1)$. We need

$$x_1^2y_1^2e + 1 = w_1^2, \tag{1}$$

$$4x_1^2y_1^2(2x_1^2 - 1)^2e + 1 = w_2^2. \tag{2}$$

Substituting (1) into (2) yields

$$4(2x_1^2 - 1)^2(w_1^2 - 1) + 1 = w_2^2,$$

so that

$$[2(2x_1^2 - 1)]^2 w_1^2 - w_2^2 = [2(2x_1^2 - 1)]^2 - 1. \qquad (3)$$

Equation (3) will be satisfied in particular when

$$2(2x_1^2 - 1)w_1 - w_2 = 1,$$
$$2(2x_1^2 - 1)w_1 + w_2 = 4(2x_1^2 - 1)^2 - 1,$$

i.e., when $w_1 = 2x_1^2 - 1$ and

$$e = \frac{w_1^2 - 1}{x_1^2 y_1^2} = \frac{4x_1^4 - 4x_1^2}{x_1^2 y_1^2} = \frac{4(x_1^2 - 1)}{y_1^2} = 4d.$$

Thus, a candidate for e is $4d$. Is it true that $4dz_n^2 + 1$ is a square for every value of n? Yes, since

$$4dz_n^2 + 1 = 4dx_n^2 y_n^2 + 1 = 4x_n^2(x_n^2 - 1) + 1 = (2x_n^2 - 1)^2.$$

Thus, with $w_n = 2x_n^2 - 1$ and $z_n = x_n y_n$, we find that $w_n^2 - 4dz_n^2 = 1$. Does this pick up all the solutions of $x^2 - 4dy^2 = 1$?

If $(x, y) = (u, v)$ is a solution of $x^2 - dy^2 = -1$, we can develop an analogous theory in which

$$x_{n+1}y_{n+1} = [4u^2 + 2]x_n y_n - x_{n-1}y_{n-1}.$$

Chapter 5

1.2(c).

$$\frac{p + r}{q + s} = \left(\frac{q}{q + s}\right)\frac{p}{q} + \left(\frac{s}{q + s}\right)\frac{r}{s}.$$

1.3(b). Yes. This is easily established by referring to the diagram in (a).

1.4(b). 47/15 is followed by a sequence in which the numerators increase by 22 and the denominators by 7 until we arrive at 333/106 and 355/113, which straddle π. The next two fractions are 688/219 and 1043/332.

1.5(e). $x^2 - 29y^2 = 1$ is satisfied by $(x, y) = (9801, 1820)$.

1.8. We know the numbers in the column headed $x^2 - dy^2$ along with the sign. Suppose we know the values of x and y yielding the numbers $x^2 - dy^2$ up to but not including the value m. Let h and p be the most recent values of the same and opposite signs, respectively, and suppose $h = r^2 - ds^2$ and $p = u^2 - dv^2$. Then $m = (r + u)^2 - d(s + v)^2$.

1.9. The successive values of (x, y) yielding the values of $x^2 - 54y^2$ are (7, 1), (8, 1), (15, 2), (22, 3), (37, 5), (59, 8), (81, 11), (103, 14), (125, 17), (147, 20), (169, 23), (316, 43), (485, 66). The last is the required solution.

1.10(a). In the list, suppose that we have got to $m = a^2 - b^2d$. With $h = r^2 - s^2d$ and $p = u^2 - v^2d$ the most recent entries of the same and opposite signs, respectively, by Exercise 1.4(e) we must have $a = u + r$ and $b = v + s$, whence $r = a - u$ and $s = b - v$.

1.10(b). To find the entry past m, we need to take note of $m = a^2 - b^2d$ and $p = u^2 - v^2d$ to get the number $(a + u)^2 - (b + v)^2d$. We can also get it directly from the table

$(a - u)^2 - (b - v)^2d$		
	$(2au - u^2) - (2bv - v^2)d$	
$a^2 - b^2d$		$2u^2 - 2v^2d$
	$(u^2 + 2au) - (v^2 + 2bv)d$	
$(a + u)^2 - (b + v)^2d$		

2.1. The common divisors are 1, 3, 9, 27.

2.3(b). Suppose, for example, that $u = x - y$ and $v = y$. If d divides x and y, then d divides their difference and so must divide u as well as v. Similarly, any divisor of u and v must divide x and y.

2.3(c). Since y is a multiple of d, $y \geq d$, so that $x - y \leq x - d$. Since $x - y$ is a multiple of d, $x - y \geq d$, so that $y \leq x - d$. Therefore, $u = \max(y, x - y) \leq x - d$. Hence as long as the pairs have distinct entries, the first entry reduces by at least $d > 0$. Since z is finite, this can happen at most finitely often.

2.4. 331.

2.7(b). If the process terminates, then $w_{k-2} = a_{k-1}w_{k-1} + w_k = (a_{k-1}a_k + 1)w_k$. Working up the set of equations in (a), we can eventually show that w and z are integer multiples of w_k, and so are commensurable. Conversely, if z and w are commensurable, then $z = mu$ and $w = nu$ for some number u and positive integers m and n. From the equations in (a), we find that $w_1, w_2, \ldots$ are successively smaller positive integer multiples of u. This cannot continue forever, and the process terminates.

2.7(c). Suppose that z and w are commensurable. Then there exists a number u and positive integers m and n for which $z = mu$ and $w = nu$, so that $z/w = m/n$ is rational. Conversely, if z/w is rational, say equal to m/n, then $z/m = w/n$, so that $z = m(z/m)$ and $w = n(z/m)$ are commensurable.

3.4(b). More generally, let $z = a_0 + 1/a_1 + 1/a_2 + \cdots + 1/a_n + 1/u_{n+1}$. We have successively

$$a_n < a_n + 1/u_{n+1},$$
$$a_{n-1} + 1/a_n > a_{n-1} + 1/a_n + 1/u_{n+1},$$
$$a_{n-2} + 1/a_{n-1} + 1/a_n < a_{n-2} + 1/a_{n-1} + 1/a_n + 1/u_{n+1},$$
$$\ldots$$
$$a_{n-2i} + 1/a_{n-2i+1} + \cdots + 1/a_n < a_{n-2i} + 1/a_{n-2i+1} + \cdots + 1/a_n + 1/u_{n+1},$$
$$a_{n-2i-1} + 1/a_{n-2i} + \cdots + 1/a_n < a_{n-2i-1} + 1/a_{n-2i} + \cdots + 1/a_n + 1/u_{n+1},$$

with the result that

$$a_0 + 1/a_1 + \cdots + 1/a_n < z$$

when n is even, and

$$a_0 + 1/a_1 + \cdots + 1/a_n > z$$

when n is odd.

3.5. For example, using the results for p_2 and q_2 with parameters a_1, a_2, a_3,

$$\begin{aligned}\frac{p_3}{q_3} &= a_0 + \frac{1}{(a_1a_2a_3 + a_1 + a_3)/(a_2a_3 + 1)}\\ &= \frac{a_0(a_1a_2a_3 + a_1 + a_3) + (a_2a_3 + 1)}{a_1a_2a_3 + a_1 + a_3}.\end{aligned}$$

Noting that $p_3 = a_0q_3 + (a_2a_3 + 1)$, $q_3 = a_1(a_2a_3 + 1) + a_3$, we see that any common divisor d of p_3 and q_3 must divide $a_2a_3 + 1$, hence a_3, hence 1, so that p_3 and q_3 must be coprime.

3.6(a). For the third equal sign, multiply numerator and denominator by a_{n+1}.

3.6(b). $\phi_n(a_0, a_1, \ldots, a_n) = \phi_n(a_n, \ldots, a_1, a_0)$ holds for $n = 0$, $n = 1$ and $n = 2$. Suppose that it has been established for indices n up to and including $m - 1 \geq 2$. Then

$$\begin{aligned}\phi_m(a_0, a_1, \ldots, a_m) &= a_m\phi_{m-1}(a_0, a_1, \ldots, a_{m-1}) + \phi_{m-2}(a_0, a_1, \ldots, a_{m-2})\\ &= a_m\phi_{m-1}(a_{m-1}, \ldots, a_1, a_0) + a_{m-2}(a_{m-2}, \ldots, a_1, a_0)\\ &= a_ma_0\phi_{m-2}(a_{m-1}, \ldots, a_1) + a_m\phi_{m-3}(a_{m-1}, \ldots, a_2)\\ &\qquad + a_0\phi_{m-3}(a_{m-2}, \ldots, a_1) + \phi_{m-4}(a_{m-2}, \ldots, a_2)\\ &= a_0[a_m\phi_{m-2}(a_1, \ldots, a_{m-1}) + \phi_{m-3}(a_1, \ldots, a_{m-2})]\\ &\qquad + [a_m\phi_{m-3}(a_2, \ldots, a_{m-1}) + \phi_{m-4}(a_2, \ldots, \phi_{m-2})]\\ &= a_0\phi_{m-1}(a_1, a_2, \ldots, a_m) + \phi_{m-2}(a_2, a_3, \ldots, a_m)\\ &= a_0\phi_{m-1}(a_m, \ldots, a_2, a_1) + \phi_{m-2}(a_m, \ldots, a_2)\\ &= \phi_m(a_m, \ldots, a_2, a_1, a_0).\end{aligned}$$

3.6(c). From (b) we can see that any common divisor of $\phi_m(a_0, \ldots, a_m)$ and $\phi_{m-1}(a_1, a_2, \ldots, a_m)$ must divide $\phi_{m-2}(a_2, \ldots, a_m)$. Continuing in this way, we see that it must divide $\phi_1(a_{m-1}, a_m) = a_{m-1}a_m + 1$ and $\phi_0(a_m) = a_m$ and so must divide 1.

3.7(a). We can use $p_{n+1}q_n - p_nq_{n+1} = a_np_nq_n + p_{n-1}q_n - a_np_nq_n - p_nq_{n-1} = -(p_nq_{n-1} - p_{n-1}q_n)$ and the fact that $p_1q_0 - p_0q_1 = 1$, $p_2q_1 - p_1q_2 = -1$.

3.7(b). It can be seen that the difference is $(-1)^n/q_nq_{n+1}$. Since $\{q_n\}$ is a strictly increasing sequence of positive integers, the result follows.

3.7(c). See the solution to Exercise 3.4(b).

3.7(e). Observe that $u_n = a_n + 1/a_{n+1} + 1/a_{n+2} + \cdots$.

3.9. After the initial negative entry for $x^2 - 29y^2$, the column for $x^2 - 29y^2$ has $a_2 = 2$ positive entries, $a_3 = 1$ negative entry, $a_4 = 1$ positive entry, $a_5 = 2$ negative entries, and so on.

3.11. Let $z = a_0 + 1/a_1 + 1/a_2 + \cdots$ and $w = b_0 + 1/b_1 + 1/b_2 + \cdots$. Suppose that for some index m, w lies between the mth and $(m + 1)$th convergents of z. This means that

$$\frac{p_m}{q_m} \leq w \leq \frac{p_{m+1}}{q_{m+1}}$$

when m is even and

$$\frac{p_m}{q_m} > w > \frac{p_{m+1}}{q_{m+1}}$$

when m is odd, where p_m/q_m is the mth convergent for z. We show that $a_i = b_i$ $(1 \leq i \leq m)$, so that the first m convergents for z and w agree.

We prove this by induction on m. For $m = 0$, the hypothesis is that

$$a_0 \leq w \leq a_0 + \frac{1}{a_1} < a_0 + 1,$$

so that $b_0 = \lfloor w \rfloor = a_0$, and the result follows. For $m = 1$, we have that

$$\begin{aligned} a_0+\frac{1}{a_1} \geq w \geq a_0+\frac{1}{a_1+1/a_2} > a_0+\frac{1}{a_1+1} &\Longrightarrow \frac{1}{a_1} \geq w-a_0 > \frac{1}{a_1+1} \\ &\Longrightarrow a_1 \leq \frac{1}{w-a_0} < a_1+1 \\ &\Longrightarrow \left\lfloor \frac{1}{w-a_0} \right\rfloor = b_1. \end{aligned}$$

Suppose that the result has been proved for all positive values of z and values of m up to $m = r - 1 \geq 0$ and that w lies between the rth and $(r + 1)$th convergents for z. Then in particular, $p_0/q_0 \leq w \leq p_1/q_1$, so that $a_0 = b_0$. Since w lies between $a_0 + 1/a_1 + \cdots + 1/a_r$ and $a_0 + 1/a_1 + \cdots + 1/a_{r+1}$, it follows that $1/(w - a_0) = 1/(w - b_0)$ lies between $a_1 + 1/a_2 + \cdots + 1/a_r$ and $a_1 + 1/a_2 + \cdots + 1/a_{r+1}$, which are the $(r - 1)$th and rth convergents of

$1/(z - a_0)$. Since $1/(w - b_0) = b_1 + 1/b_2 + \cdots + 1/b_r$, it follows from the induction hypothesis that $a_1 = b_1, \ldots, a_r = b_r$.

3.12(a). Solve the system to get fractions x and y whose denominators are $p_{n+1}q_n - p_n q_{n+1} = (-1)^n$.

3.12(b). If $x = 0$, then $b = q_{n+1}y$, whence $y > 0$ and $b \geq q_{n+1}$, contrary to hypothesis.

3.12(c). If $y = 0$, then $|zb - a| = |x||q_n z - p_n| \geq |q_n z - p_n|$, contrary to hypothesis.

3.12(d) Suppose $y < 0$; then $b < q_n x$ so $x > 0$. Suppose $y > 0$; then $b < q_{n+1} \leq y q_{n+1}$, so that $q_n x$, and hence x, are negative.

3.12(e). Note that $q_n z - p_n$ and $q_{n+1}z - p_{n+1}$ have opposite signs, as do x and y.

3.13. The hypothesis yields $|bz - a| < \frac{b}{q_n}|q_n z - p_n|$. If $b \leq q_n$, then by Exercise 3.12, we must have $b \geq q_{n+1}$, which yields a contradiction, since $q_n < q_{n+1}$.

4.1(c). Note that $u_r^2 p_{r-1}q_{r+h-1} + u_r(p_{r-1}q_{r+h-2} + p_{r-2}q_{r+h-1}) + p_{r-2}q_{r+h-2} = u_r^2 q_{r-1}p_{r+h-1} + \cdots$ and that p_{r-1}/q_{r-1} and p_{r+h-1}/q_{r+h-1} are distinct fractions, each in lowest terms.

4.1(d). Since z is a root of

$$0 = (z - p + \sqrt{q})(z - p - \sqrt{q}) = z^2 - 2pz + (p^2 - q),$$

z is a root of a quadratic equation with rational coefficients. Multiply by the least common multiple of the denominator of the coefficients to get the result.

4.2(a). Since

$$z = \frac{u_n p_{n-1} + p_{n-2}}{u_n q_{n-1} + q_{n-2}},$$

we find that

$$A(u_n p_{n-1} + p_{n-2})^2 + B(u_n p_{n-1} + p_{n-2})(u_n q_{n-1} + q_{n-2}) + C(u_n q_{n-1} + q_{n-2})^2 = 0,$$

from which the result follows.

4.2(b). Since z is irrational, so is u_n. Hence, u_n cannot be a root of a linear equation with integer coefficients, so A_n must be nonzero.

4.2(d).

$$\begin{aligned} A_n &= A_n - (Az^2 + Bz + C)q_{n-1}^2 \\ &= A(p_{n-1} - zq_{n-1})(p_{n-1} + zq_{n-1}) + B(p_{n-1} - zq_{n-1})q_{n-1} \\ &= A(p_{n-1} - zq_{n-1})(p_{n-1} - zq_{n-1} + 2zq_{n-1}) + B(p_{n-1} - zq_{n-1})q_{n-1}. \end{aligned}$$

Since $|p_{n-1} - zq_{n-1}| < 1/q_{n-1} < 1$, we find that

$$|A_n| < |A|\left(\frac{1}{q_{n-1}}\right)|1 + 2zq_{n-1}| + |B|\left(\frac{1}{q_{n-1}}\right)q_{n-1} \leq |A| + 2|Az| + |B|.$$

The inequality for C_n is similar to that for A_n, and the inequality for B_n is a consequence of $B_n^2 = 4A_nC_n + B^2 - 4AC$.

4.3(a). Suppose that $Az^2 + Bz + C = 0$. Then $z = (-B \pm \sqrt{B^2 - 4AC})/2A$. Let $d = B^2 - 4AC$, $k_0 = \mp B$, and $h_0 = \pm 2A$, depending on the sign before the surd. The two roots of the quadratic are $(k_0 + \sqrt{d})/h_0$ and $(k_0 - \sqrt{d})/h_0 = (-k_0 + \sqrt{d})/(-h_0)$; their product is $(k_0^2 - d)/h_0^2 = C/A$. Thus

$$\frac{d - k_0^2}{h_0} = \frac{-Ch_0}{A} = \mp 2C$$

is an integer. For example, the roots of the quadratic $7x^2 - 8x + 2 = 0$ are $(4 + \sqrt{2})/7$ and $(-4 + \sqrt{2})/(-7)$.

4.3(b).

$$\frac{1}{u_1} = \frac{k_0 + \sqrt{d} - h_0 a_0}{h_0} = \frac{(k_0 - h_0 a_0)^2 - d}{[(k_0 - h_0 a_0) - \sqrt{d}]h_0},$$

from which the result follows:

$$h_1 = \left(\frac{d - k_0^2}{h_0}\right) + 2a_0 k_0 - a_0^2 h_0$$

and $k_1 = -(k_0 - a_0 h_0)$. Observe that $d - k_1^2 = d - (k_0 - a_0 h_0)^2 = h_0 h_1$.

4.3(d). For example, $z = (4 + \sqrt{2})/7$. Then $a_0 = 0$, $u_1 = (-4 + \sqrt{2})/(-2)$, $a_1 = 1$, $u_2 = 2 + \sqrt{2}$, $a_2 = 3$, and for $i \geq 3$, $u_i = 1 + \sqrt{2}$ and $a_i = 2$.

4.4(a). The mapping $g(x + y\sqrt{d}) = x - y\sqrt{d}$, with x and y rational, satisfies $g(\alpha \pm \beta) = g(\alpha) \pm g(\beta)$, $g(\alpha\beta) = g(\alpha)g(\beta)$, and $g(\alpha/\beta) = g(\alpha)/g(\beta)$. Since p_i and q_i are rational, we find that

$$w = g(z) = g\left(\frac{u_n p_{n-1} + p_{n-2}}{u_n q_{n-1} + q_{n-2}}\right) = \frac{g(u_n p_{n-1} + p_{n-2})}{g(u_n q_{n-1} + q_{n-2})}$$
$$= \frac{g(u_n)p_{n-1} + p_{n-2}}{g(u_n)q_{n-1} + q_{n-2}} = \frac{v_n p_{n-1} + p_{n-2}}{v_n q_{n-1} + g_{n-2}},$$

as desired.

4.4(b). Solve the equation in (a) for v_n. By Exercise 3.7, $\lim_{k\to\infty} p_n/q_n = z$, so that the quantity in parentheses is close to the positive value 1 when n is large. Since all q_i are positive, the result follows.

4.4(c). Note that $u_n - v_n = 2\sqrt{d}/h_n$.

4.4(d). Since $k_n^2 < d$ for $n > N$, k_n can assume at most $2d + 1$ values. For each value of k_n there are only finitely many pairs of integers whose products are $d - k_n^2$.

4.5. $f(x) = q_{s-1}x^2 + (q_{s-2} - p_{s-1})x - p_{s-2}$. The equation $f(x) = 0$ has two roots, one of which is z. Any quadratic equation with integer coefficients satisfied by z must be satisfied by w, so that w is the second root of the equation. Since $zw = -p_{s-2}/q_{s-1}$, w must be negative. Since $f(0) = -p_{s-2} < 0$ and $f(-1) = (p_{s-1} - p_{s-2}) + (q_{s-1} - q_{s-2}) > 0$, $f(x)$ must vanish for some value of x between -1 and 0; this must be w. Observe that for any continued fraction, each a_i for $i \geq 1$ must be a positive integer, so that since a_0 appears later, $a_0 \geq 1$.

4.6(a). We have that $u_i = a_i + 1/u_{i+1}$. Passing to the surd conjugate yields the result. Since $-1 < w = a_0 + 1/v_1 < 0$, it follows that $-1 - a_0 < 1/v_i < -a_0 \le -1$, so that v_i must lie between -1 and 0. Suppose it has been shown that $-1 < v_i = a_i + 1/v_{i+1} < 0$. Then the same argument yields $-1 < v_{i+1} < 0$.

4.6(d). If $z = \sqrt{d} + \lfloor\sqrt{d}\rfloor$, then $w = \lfloor\sqrt{d}\rfloor - \sqrt{d}$; z and w satisfy $-1 < w < 0 < 1 < z$.

4.7(b). We have that $A + B\sqrt{d} = 0$, where

$$A = dq_n - k_{n+1}p_n - h_{n+1}p_{n-1},$$
$$B = k_{n+1}q_n + h_{n+1}q_{n-1} - p_n.$$

Since A and B are integers and $\sqrt{d}$ is irrational, it follows that $A = B = 0$ or

$$p_n = k_{n+1}q_n + h_{n+1}q_{n-1},$$
$$dq_n = k_{n+1}p_n + h_{n+1}p_{n-1}.$$

Hence $p_n^2 - dq_n^2 = h_{n+1}(p_nq_{n-1} - p_{n-1}q_n) = h_{n+1}(-1)^{n-1}$ by Exercise 3.7(a).

4.7(c). $h_{ns} = h_0 = 1$ for each positive integer n.

5.1. The first three convergents are t, $(2t^2 + 1)/2t$, $(4t^3 + 3t)/(4t^2 + 1)$.

5.2(a). We find that

$$\sqrt{t^2 + c} = t + \frac{c}{\sqrt{t^2 + c} + t},$$
$$\frac{\sqrt{t^2 + c} + t}{c} = \frac{2t}{c} + \frac{1}{\sqrt{t^2 + c} + t},$$

so that

$$\sqrt{t^2 + c} = t + 1/(2t/c) + 1/2t + 1/(2t/c) + 1/2t + \cdots.$$

The convergents are p_n/q_n as given in the following table:

n	p_n	q_n	$p_n^2 - (t^2 + c)q_n$
1	t	1	$-c$
2	$2t^2 + c$	$2t$	c^2
3	$4t^3 + 3ct$	$4t^2 + c$	$-c^3$
4	$8t^4 + 8ct^2 + c^2$	$8t^3 + 4ct$	c^4

Dividing p_{2m} and q_{2m} by c^m yields solutions (x, y) for $x^2 - (t^2 + c)y^2 = 1$.

5.2(b). When $c = 2$, we find that

$$\sqrt{t^2 + 2} = t + 1/t + 1/2t + 1/t + 1/2t + \cdots,$$

so that in place of the table of (a) we have that

n	p_n	q_n	$p_n^2 - (t^2+2)q_n^2$
1	t	1	-2
2	t^2+1	t	1
3	$2t^3+3t$	$2t^2+1$	-2
4	$2t^4+4t^2+1$	$2t^3+2t$	1

$(x, y) = (t^2+1, t)$ satisfies $x^2 - (t^2+2)y^2 = 1$.

5.3. The convergents are p_n/q_n as given in the following table:

n	p_n	q_n	$p_n^2 - (t^2-1)q_n^2$
1	$t-1$	1	$2(1-t)$
2	t	1	1
3	$2t^2-t-1$	$2t-1$	$2(1-t)$
4	$2t^2-1$	$2t$	1

5.4. We find that

$$\sqrt{4t^2+12t+5} = (2t+2) + \frac{4t+1}{\sqrt{4t^2+12t+5}+(2t+2)},$$

$$\frac{\sqrt{4t^2+12t+5}+(2t+2)}{4t+1} = 1 + \frac{4}{\sqrt{4t^2+12t+5}+(2t-1)},$$

$$\frac{\sqrt{4t^2+12t+5}+(2t-1)}{4} = t + \frac{2t+1}{\sqrt{4t^2+12t+5}+(2t+1)},$$

and so on. Eventually,

$$\sqrt{4t^2+12t+5} = (2t+2) + 1/1 + 1/t + 1/2 + 1/t + 1/1 + 1/2(2t+2) + 1/1 + \cdots,$$

with periodicity. The convergent

$$(2t+2) + 1/1 + 1/t + 1/2 + 1/t + 1/1$$

is equal to $(4t^3+18t^2+24t+9)/(2t^2+6t+4)$, and we find that

$$(4t^3+18t^2+24t+9)^2 - (4t^2+12t+5)(2t^2+6t+4)^2 = 1.$$

5.5(b). $(x, y) = (mt^3+nt^2+1, t)$.

5.6(a). $(x, y) = (4t^3+6t^2+6t+2, 2t^2+2t+1)$.

5.6(b). $(x, y) = (4t^4-4t^3+2t^2+2t-1, 2t^2-2t+1)$.

5.6(c). $(x, y) = (16t^4+4t^3+8t^2+3t+1, 4t^2+1)$.

5.7(b).

$$\sqrt{k^2s^2+k} = ks + \frac{k}{\sqrt{k^2s^2+k}+ks},$$
$$\frac{\sqrt{k^2s^2+k}+ks}{k} = 2s + \frac{1}{\sqrt{k^2s^2+k}+ks},$$

leads to

$$\sqrt{k^2s^2+k} = ks + 1/2s + 1/2ks + 1/2s + 1/2ks + \cdots,$$

with convergents

$$ks, \quad \frac{2ks^2+1}{2s}, \quad \frac{4k^2s^3+3ks}{4ks^2+1}, \quad \ldots .$$

A solution of $x^2 - (k^2s^2+k)y^2 = 1$ is

$$(x, y) = (2ks^2+1, 2s).$$

5.7(c). $d = 2, 5, 6, 10, 12, 17, 18, 20, 26, 37, 38, 39, 50, 66, 68, 84, \ldots .$

5.8(a).

s	0	1	2	3	4	5	6
t	2	5	8	11	14	17	20
t^2+3	7	28	67	124	199	292	403
a_0	2	5	8	11	14	17	20
a_1	1	3	5	7	9	11	13
a_2	1	2	2	2	2	2	2
a_3	1	3	1	1	1	1	1
a_4	4	10	1	1	2	3	3
a_5	1	3	7	1	2	8	1
a_6	1	2	1	3	5	3	3
a_7	1	3	1	1	4	1	1
a_8	4	10	2	4	1	2	2
a_9	1	3	5	1	1	11	13
a_{10}	1	2	16	3	13	34	40

5.8(b).

$$\sqrt{9s^2+12s+7} = (3s+2) + \frac{3}{\sqrt{9s^2+12s+7}+(3s+2)},$$
$$\frac{\sqrt{9s^2+12s+7}+(3s+2)}{3} = (2s+1) + \frac{2(s+1)}{\sqrt{9s^2+12s+7}+(3s+1)},$$
$$\frac{\sqrt{9s^2+12s+7}+(3s+1)}{2(s+1)} = 2 + \frac{\sqrt{9s^2+12s+7}-(s+3)}{2(s+1)}$$
$$= 2 + \frac{4s-1}{\sqrt{9s^2+12s+7}+(s+3)},$$

$$\frac{\sqrt{9s^2+12s+7}+(s+3)}{4s-1} = 1 + \frac{\sqrt{9s^2+12s+7}-(3s-4)}{4s-1}$$
$$= 1 + \frac{9}{\sqrt{9s^2+12s+7}+(3s-4)}.$$

When $s = 0$, the quantity on the left in the third equation is less than 2; when $s = 1$, the quantity on the left in the fourth equation exceeds 3.

5.8(c),

n	a_n	p_n	q_n	$p_n^2 - dq_n^2$
0	$3s+2$	$3s+2$	1	-3
1	$2s+1$	$6s^2+7s+3$	$2s+1$	$2(s+1)$
2	2	$12s^2+17s+8$	$4s+3$	$-4s+1$
3	1	$18s^2+24s+11$	$6s+4$	9

The desired solution is $(x, y) = (6s^2 + 8s + \frac{11}{3}, 2s + \frac{4}{3})$.

5.9. We have that $(3s+1)^2 - (9s^2+6s+4) = -3$. Since

$$(3s+1+\sqrt{9s^2+6s+4})^2 = (18s^2+12s+5) + (6s+2)\sqrt{9s^2+6s+4},$$

we find that $(18s^2+12s+5)^2 - (9s^2+6s+4)(6s+2)^2 = 9$. The desired solution is $(x, y) = (6s^2+4s+\frac{5}{3}, 2s+\frac{2}{3})$.

5.11.

$$\sqrt{4s^2+4s-3} = 2s + 1/1 + 1/(s-1) + 1/2 + 1/(s-1) + 1/1 + 1/4s + \cdots.$$

The convergents yields the relations

$$(2s+1)^2 - (4s^2+4s-3) = (4s^2+4s-1)^2 - (4s^2+4s-3)(2s+1)^2 = 4$$

and

$$(4s^3+6s^2-1)^2 - (4s^2+4s-3)(2s^2+2s)^2 = 1.$$

5.12.

$$\sqrt{4s^2-4} = (2s-1) + 1/1 + 1/(s-2) + 1/1 + 1/2(2s-1) + 1/1 + \cdots.$$

The convergents yield the relations

$$(2s)^2 - (4s^2-4)\cdot 1 = 4,$$
$$(2s^2-1)^2 - (4s^2-4)\cdot s^2 = 1.$$

For the numerical substitution $s = 2$, the continued fraction development given above is not valid; rather

$$\sqrt{12} = 3 + 1/2 + 1/6 + 1/2 + \cdots.$$

6.1(a). Since $r^2 - ds^2 > 0$, clearly $r > s\sqrt{d}$. It follows that

$$0 < r - s\sqrt{d} = \frac{a}{r+s\sqrt{d}} < \frac{\sqrt{d}}{2s\sqrt{d}} = \frac{1}{2s}.$$

6.1(b). Similarly, $s\sqrt{d} > r$ and

$$0 < s - \frac{r}{\sqrt{d}} = \frac{s\sqrt{d} - r}{\sqrt{d}} = \frac{ds^2 - r^2}{\sqrt{d}(s\sqrt{d} + r)} < \frac{b}{2r\sqrt{d}} < \frac{1}{2r}.$$

6.2. We have $1/\sqrt{d} = 0 + 1/\sqrt{d}$, from which it is clear that $1/\sqrt{d} = 0 + 1/a_0 + 1/a_2 + \cdots$, and the convergents are the reciprocals of the convergents for $\sqrt{d}$.

6.3(a). Prove by contradiction.

6.3(c). $|(p_n/q_n) - (r/s)| \le |(p_n/q_n) - \sqrt{d}| + |\sqrt{d} - (r/s)|$.

6.3(e). Let $r = gu$ and $s = gv$, where g is the greatest common divisor of r and s. Then

$$\left|\sqrt{d} - (u/v)\right| = \left|\sqrt{d} - (r/s)\right| < \frac{1}{2s^2} \le \frac{1}{2v^2},$$

whence $u/v = p_n/q_n$.

6.5. If there is a solution (x, y) for which the greatest common divisor of x and y is 1, then by Exercises 6.2, 6.3, 6.4, $(x, y) = (p_n, q_n)$ for some n. If the greatest common divisor of a solution is g, then k is a multiple of g^2, i.e., $k = g^2 l$. We have $l < k < \sqrt{d}$, and so we can find $(x, y) = (gp_n, gq_n)$ for some n.

Chapter 6

1.1. $\{t_n\} = \{1, 6, 35, 204, 1189, 6930, 40391, \ldots\}$.

1.2. Use the fact that $t_{n+1} + t_n = 7(t_n + t_{n-1}) - 8t_{n-1}$ and an induction argument.

1.3. $\{(a_n, b_n, c_n\} = \{(3, 4, 5), (20, 21, 29), (119, 120, 169), \ldots\}$.

1.5(a). For $n \ge 4$,

$$\begin{aligned} p_n &= 2p_{n-1} + p_{n-2} = 2(2p_{n-2} + p_{n-3}) + p_{n-2} = 5p_{n-2} + (p_{n-2} - p_{n-4}) \\ &= 6p_{n-2} - p_{n-4}, \end{aligned}$$

with a similar result for q_n.

1.5(b). The sequences $\{p_{2n+1}\}, \{q_{2n+1}\}, \{c_n = t_{n+1} - t_n\}$, and $\{a_n + b_n = t_{n+1} + t_n\}$ all satisfy the recursion $x_n = 6x_{n-1} - x_{n-2}$ for $n \ge 3$. The desired result follows by induction.

2.16. $(x, y) = (17, -7), (8, -2), (7, 1), (13, 5), (32, 14)$.

3.2–3.3 Let $g = v - u$, so that $g(r + s) = \alpha b + (\beta - 1)a$, so that $v = u + g$. Then (5) becomes

$$\begin{aligned} &ru^2 + s(u + e)^2 + 2g[ru + s(u + e)] + (r + s)g^2 = (\alpha b + \beta a - b)^2 \\ &\implies (b - a)^2 + 2g(a + b) + kg^2 = (\alpha b + \beta a - b)^2. \end{aligned}$$

Multiplying by k yields

$$k(b-a)^2+2[\alpha b+(\beta-1)a](a+b)+[\alpha b+(\beta-1)a]^2=k[\alpha b+\beta a-b]^2$$
$$\Rightarrow [\alpha b+(\beta-1)a]^2+2[\alpha b+(\beta-1)a](a+b)$$
$$=k[(\alpha^2-2\alpha)b^2+2(\alpha\beta-\beta+1)ab+(\beta^2-1)a^2]$$
$$\Rightarrow (\alpha^2+2\alpha)b^2+2[\alpha\beta+\beta-1]ab+(\beta^2-1)a^2$$
$$=k(\alpha^2-2\alpha)b^2+2k(\alpha\beta-\beta+1)ab+k(\beta^2-1)a^2.$$

3.4(a). Subtract (2) from (1) to obtain $(r+s)(u-v)=2a$. Since $r+s=k=-1$, $v=2a+u$. For equation (4), we check that

$$r(u+2a)^2+s(u+2a+e)^2$$
$$=ru^2+s(u+e)^2+4a[ru+s(u+e)]+(r+s)4a^2$$
$$=(b-a)^2+4a(a+b)-4a^2=(b+a)^2.$$

3.7. $\{t_n\}=\{\ldots,-682,-117,-20,-3,2,15,88,513,2990,\ldots\}$ gives rise to the Pythagorean triples $(-403,-396,565)$, $(-72,-65,97)$, $(-15,-8,17)$, $(-4,3,5)$, $(48,55,73)$, $(297,304,425)$, $(1748,1755,2477)$, among others.

3.10(a). $(r,s,k,\alpha)=(1,2,3,4)$.

3.10(b). If we want $y=x+1$, we use the same sequence $\{t_n\}$, but the successive values of (x,y,z) are $(1,2,3)$, $(17/3,20/3,11)$, $(23,24,41)$, $(263/3,266/3,153)$. For $y=x-1$, the successive values of (x,y,z) are $(7/3,4/3,3)$, $(7,6,11)$, $(73/3,70/3,41)$, $(89,88,153)$.

3.12. (47, 65, 122).

3.13. (96, 101, 155).

4.1(c). We are led to $12f-5g=\pm1$ and the possibilities

$$(f,g;d,q,x,y)=(2,5;29,70,9801,1820),(3,7;58,99,19603,2574),$$
$$(7,17;338,239,114243,6214),$$
$$(8,9;425,268,143649,6968).$$

4.2(c). Use $3^2=1^2+2\cdot2^2$, and let $b=1$, $c=2$.

f	1	1	2	2	3	3	4
g	1	3	3	5	5	7	7
q	5	13	14	22	23	31	32
d	3	19	22	54	59	107	114
x	26	170	197	485	530	962	1025
y	15	39	42	66	69	93	96

4.2(d). Use $9^2=7^2+2\cdot4^2$ and $11^2=7^2+2\cdot6^2$.

Chapter 7

1.1. $x^2 + rxy + r^2y^2 = \frac{1}{2}[x^2 + (x + ry)^2 + r^2y^2]$ is a positive divisor of ± 1 and so must equal 1. Two of the squares in the sum must take the value 1 and the third square the value 0. If $r \neq \pm 1$, the only possibilities are $(x, y) = (\pm 1, 0)$. If $r = \pm 1$, then we also have $(x, y) = (0, \pm 1)$ as possible solutions.

1.3(a). Suppose $x^3 - dy^3 = 1$ with $y = 2$. Then $x^3 \equiv 1 \pmod 8$, which occurs if and only if $x = 8u + 1$ for some integer u. In this case

$$d = \frac{1}{8}\left[(8u + 1)^3 - 1\right] = u\left(64u^2 + 24u + 3\right).$$

1.3(b). When $u = -1$, $d = -43$, and we find that $(-7)^3 - (-43)2^3 = 1$, which in turn yields $(-7)^3 - 43(-2)^3 = 1$.

1.4(a). Suppose $x^3 - dy^3 = 1$ with $y = 3$. Then $x^3 \equiv 1 \pmod{27}$, which occurs if and only if $x \equiv 1 \pmod 9$. If $x = 9u + 1$, then $d = u(27u^2 + 9u + 1)$.

1.4(b). Take $u = -1$ and $u = +1$ to obtain solutions for the cases $d = 19$ and $d = 37$.

1.5(a). Note that $x^2 + x + 1 = (x - 1)(x + 2) + 3$.

1.5(b). $x^2 + x + 1 \equiv 0 \pmod 9 \Leftrightarrow (2x + 1)^2 + 3 \equiv 0 \pmod 9 \Leftrightarrow -3$ is a square modulo 9. But the only squares, modulo 9, are 0, 1, 4, 7.

1.6(b). (18, 7).

1.8. Values of $(d; x, y)$ for which $x^3 - dy^3 = 1$ are (7; 2, 1), (9; −2, −1), (17; 18, 7), (19; −8, −3), (26; 3, 1), (28; −3, −1), (37; 10, 3), (43; −7, −2), (63; 4, 1), (65; −4, −1), (91; 9, 2).

2.4(c). Let $r + s\omega$ be an algebraic integer and a unit. The norm of $r + s\omega$, namely $r^2 - rs + s^2$, must be equal to 1. The quantity $r + s\omega$ and its complex conjugate $r + s\omega^2$ are roots of the quadratic equation

$$0 = (x - r - s\omega)(x - r - s\omega^2) = x^2 - (2r - s)t + (r^2 - rs + s^2).$$

Hence $2r - s$ is an integer, so $3s^2 = 4 - (2r - s)^2$ is an integer. Thus, s is an integer. Since r is a rational root of the quadratic equation $x^2 - sx + (s^2 - 1) = 0$, r must be an integer.

2.5(a). Suppose $\rho = \sigma\tau$, where neither σ nor τ is a unit. Then $N(\rho) = N(\sigma)N(\tau)$, so that $N(\rho)$ is an integer with two nontrivial integer factors. The result follows.

2.5(b). $N(1 - \omega) = 3$.

2.8(b). $N(\delta) = N(\alpha - \beta\gamma) = N(\beta)N((\alpha/\beta) - \gamma) = N(\beta)N((u - m) + (v - m)\omega)$.

3.1(a). For example, $(x - 1) - (x - \omega^2) = -(1 + \omega)(1 - \omega) = \omega^2(1 - \omega)$.

3.1(f). Since $(1 - \omega)^{-1}(x - 1)$, $(1 - \omega)^{-1}(x - \omega)$, and $(1 - \omega)^{-1}(x - \omega^2)$ are pairwise relatively coprime algebraic integers whose product is a perfect cube, by Exercise 2.12(d) each must be a cube up to a unit factor.

3.2(b). Let $\gamma_1 = r_1 + s_1\omega = (r_1 + s_1) - s_1(1 - \omega)$, so that $\gamma_1 \equiv r_1 + s_1 \pmod{(1 - \omega)}$. Since $r_1 + s_1$ is not divisible by $1 - \omega$, it is not divisible by $3 = (1 - \omega)(1 - \omega^2)$. Hence $r_1 + s_1 \equiv \pm 1 \pmod 3$ (with respect to the usual integers), and so $r_1 + s_1 \equiv \pm 1 \pmod{(1 - \omega)}$. A similar result holds for γ_2.

4.8(b). Since

$$(x + ay + a^2z)[(x - ay)^2 + a^2(y - az)^2 + (a^2z - x)^2] = 2$$

and the terms of the second factor on the left can be only 0, 1, or 2, then $a^2(y - az)^2 = 0$, so that $x - ay = x - a^2z$. Hence the second factor is $2(x - a^2z)$. Since this cannot vanish, we must have $x = a^2z \pm 1$, so that the first factor is $\pm 1 + 3a^2z$. The plus sign must be taken, and so $(x, y, z) = (1, 0, 0)$.

4.8(c). In the case $a = \pm 1$, if any two terms of the second factor vanish, then the third must as well. Hence two terms of the second factor must be 1, and the third one 0. Suppose that $x = z$. Then $y - az = \pm 1$, so that the first factor must be $1 = 3z + a$. This leads to $(a; x, y, z) = (\pm 1; 0, \pm 1, 0)$. Suppose $x = ay$. Then $y - az = \pm 1$, so that $a^2z = ay \mp a$, and the first factor must be $1 = 3ay \mp a$, which leads to $\mp a = 1$ and $(x, y, z) = (0, 0, 1)$. Finally, suppose that $y = az$. Then $x - ay = x - a^2z = \pm 1$, so the first factor must be $3a^2z \pm 1 = 1$, where the plus sign must be taken and $(x, y, z) = (1, 0, 0)$. Thus, the solutions are $(x, y, z) = (1, 0, 0), (0, a, 0), (0, 0, 1)$.

5.1(c). Use Exercise 4.5.

5.2(c). Clearly, $\alpha + \tau_1(\alpha) + \tau_2(\alpha) = 3u$ and $\alpha\tau_1(\alpha)\tau_2(\alpha) = u^3 + cv^3 + c^2w^3 - 3cuvw$.

$$\begin{aligned}
\alpha(\tau_1(\alpha) &+ \tau_2(\alpha)) + \tau_1(\alpha)\tau_2(\alpha) \\
&= (u + v\theta + w\theta^2)(2u - v\theta - w\theta^2) \\
&\quad + (u^2 - cvw) + (cw^2 - uv)\theta + (v^2 - uw)\theta^2 \\
&= (2u^2 - 2cvw) + (uv - cw^2)\theta + (uw - v^2)\theta^2 \\
&\quad + (u^2 - cvw) + (cw^2 - uv)\theta + (v^2 - uw)\theta^2 \\
&= 3(u^2 - cvw).
\end{aligned}$$

The desired equation can be determined from

$$0 = (t - \alpha)(t - \tau_1(\alpha))(t - \tau_2(\alpha)).$$

5.3. Since ϵ is real, $\epsilon = \pm 1$. $N(1) = 1$ and $N(-1) = -1$.

5.4(a) Since $|\epsilon||\tau_1(\epsilon)|^2 = 1$ and $1 \leq |\epsilon| \leq M$, it follows that $M^{-\frac{1}{2}} \leq |\tau_i(\epsilon)| \leq 1$ for $i = 1, 2$. If $\epsilon = u + v\theta + w\theta^2$, then

$$|3u| = |\epsilon + \tau_1(\epsilon) + \tau_2(\epsilon)| \leq M + 2,$$

and since $-cvw = (u^2 - cvw) - u^2$,

$$\begin{aligned}|3cvw| &\le |3(u^2 - cvw)| + |3u^2| \\ &\le |\epsilon||\tau_1(\epsilon)| + |\epsilon||\tau_2(\epsilon)| + |\tau_1(\epsilon)\tau_2(\epsilon)| + 3u^2 \\ &\le 2M + 1 + \frac{1}{3}(M+2)^2 = \frac{1}{3}(M^2 + 10M + 7) \le 6M^2.\end{aligned}$$

There are only finitely many possible choices of the integer u, and for each choice of u, finitely many pairs of integers (v, w) for which $|vw| \le 2M^2/c$. Thus there are only finitely many units $\epsilon = u + v\theta + w\theta^2$ with $|\epsilon| \le M$.

5.4(b). Let $\epsilon \in E$ and $|\epsilon| \ne 1$. Then one of the elements $\epsilon, -\epsilon, 1/\epsilon, -1/\epsilon$ belongs to E and exceeds 1; call it δ. There are finitely many elements in E that lie between 1 and δ; let γ be the smallest of these.

5.4(c). Let $\delta \in E$ with $\delta > 0$. Select that integer n for which $\gamma^m \le \delta < \gamma^{m+1}$. Then $\gamma^m\delta^{-1}$ is a unit that is not less than 1 but strictly less than γ. Since γ is minimal, $\gamma\delta^{-1} = 1$ so $\delta = \gamma$.

5.5(d). Note that $|\theta|$ and $|\theta^2|$ do not exceed c and that $|\omega - 1|$ and $|\omega^2 - 1|$ do not exceed 2.

5.5(f). Given any finite collection of elements $u + v\theta + w\theta^2$ whose norms do not exceed $25c^2$ in absolute value, we can select n such that $4n^2$ exceeds the largest value $|u + v\theta + w\theta^2|^{-1}$. Now apply (c) and (d) to obtain a new element.

5.6(b). We have that

$$\frac{m}{u_2 + v_2\theta + w_2\theta^2} = (u_2 - cv_2w_2) + (cw_2^2 - u_2v_2)\theta + (v_2^2 - u_2w_2)\theta^2,$$

whereupon

$$\begin{aligned}&u_3 + v_3\theta + w_3\theta^2 \\ &\quad = [u_1(u_2^2 - cv_2w_2) + cv_1(v_2^2 - u_2w_2) + cw_1(cw_2^2 - u_2v_2)] \\ &\qquad + [u_1(cw_2^2 - u_2v_2) + v_1(u_2^2 - cv_2w_2) + cw_1(v_2^2 - u_2w_2)]\theta \\ &\qquad + [u_1(v_2^2 - u_2w_2) + v_1(cw_2^2 - u_2v_2) + w_1(u_2^2 - cv_2w_2)]\theta^2 \\ &\quad \equiv [u_1^3 + cv_1^3 + c^2w_1^3 - 3cu_1v_1w_1] + 0 \times \theta + 0 \times \theta^2 = m,\end{aligned}$$

modulo m.

6.4. $(1, 1, 1)^{-1} = (-1, 1, 0)$. The recursion gives

$$\begin{aligned}(1,1,1)^2 &= 3(1,1,1) + 3(1,0,0) + (-1,1,0) = (5,4,3), \\ (1,1,1)^3 &= 3(5,4,3) + 3(1,1,1) + (1,0,0) = (19,15,12).\end{aligned}$$

6.5(b). $(x, y, z) = (4, 3, 2), (-2, 0, 1)$ both work.

7.1. We can try $y = -rz$, whereupon $cz^3 = r^3z^3 - 3rz^2$, so $3r = (r^3 - c)z$. Checking out factorizations of $3r$ leads to possibilities such as

$$(c; x, y, z) = (7; 1, -12, 6), (9; 1, 12, -6), (26; 1, -27, 9), (28; 1, 27, -9).$$

Another possibility is to let $(y, z) = (-rs, r^2)$, whereupon $c = r^{-3}s(s^2 - 3)$. Select (r, s) to make this an integer. For example, $(r, s) = (11, 753)$ yields $(c; x, y, z) = (320778; 1, -8283, 121)$.

Solutions can be found for $c = 2, 4, 5, 6, 10, 11, 14, 18, 24, 30, 36, 52, 58, 61, 67, 70, 76$. These can be found in the list for Exercise 7.3(c) and 7.3(d).

7.3(c). Note that we can use Exercise 7.2 to obtain solutions for positive values of c from those for $-c$.

c	r	t	(x, y, z)	$(x, y, z)^{-1}$
2	1	1	(1, 3, −3)	(19, 15, 12)
6	−2	−1	(1, −6, 3)	(109, 60, 33)
7	−1	−2	(1, −12, 6)	(505, 264, 138)
9	1	2	(1, 12, −6)	(649, 312, 150)
10	2	1	(1, 6, −3)	(181, 84, 39)
24	−3	−1	(1, −9, 3)	(649, 225, 78)
26	−1	−3	(1, −27, 9)	(6319, 2133, 720)
28	1	3	(1, 27, −9)	(6805, 2241, 738)
30	3	1	(1, 9, −3)	(811, 261, 84)
60	−4	−1	(1 − 12, 3)	(2161, 552, 141)
62	−2	−2	(1, −24, 6)	(8929, 2256, 570)
63	−1	−4	(1, −48, 12)	(36289, 9120, 2292)
65	1	4	(1, 48, −12)	(37441, 9312, 2316)
66	2	2	(1, 24, −6)	(9505, 2352, 582)
68	4	1	(1, 12, −3)	(2449, 600, 147)
120	−5	−1	(1, −15, 3)	(5401, 1095, 222)
124	−1	−5	(1, −75, 15)	(139501, 27975, 5610)
126	1	5	(1, 75, −15)	(141751, 28275, 5640)
130	5	1	(1, 15, −3)	(5851, 1155, 228)
213	−3	−2	(1, −36, 6)	(46009, 7704, 1290)
214	−2	−3	(1, −54, 9)	(104005, 17388, 2907)
215	−1	−6	(1, −108, 18)	(417961, 69768, 11646)
217	1	6	(1, 108, −18)	(421849, 70200, 11682)
218	2	3	(1, 54, −9)	(105949, 17604, 2925)
219	3	2	(1, 36, −6)	(47305, 7848, 1302)

7.3(d). Let $r = 3t$ so that $k = ts$. Then $c = t^3s^3 + 3t$ and $(x, y, z) = (1, ts^2, -s)$.

c	t	s	(x, y, z)	$(x, y, z)^{-1}$
2	−2	−1	(1, −2, 1)	(5, 4, 3)
4	1	1	(1, 1, −1)	(5, 3, 2)
5	−1	−2	(1, −4, 2)	(41, 24, 14)
11	1	2	(1, 4, −2)	(89, 40, 18)
14	2	1	(1, 2, −1)	(29, 12, 5)
18	−3	−1	(1, −3, 1)	(55, 21, 8)
30	1	3	(1, 9, −3)	(811, 261, 84)
36	3	1	(1, 3, −1)	(109, 33, 10)
52	−4	−1	(1, −4, 1)	(209, 56, 15)
58	−2	−2	(1, −8, 2)	(929, 240, 62)
61	1	−4	(1 − 16, 4)	(3905, 992, 252)
67	1	4	(1, 16, −4)	(4289, 1056, 260)
70	2	2	(1, 8, −2)	(1121, 272, 66)
76	4	1	(1, 4, −1)	(305, 72, 17)
110	−5	−1	(1, −5, 1)	(551, 115, 24)
122	−1	−5	(1, −25, 5)	(15251, 3075, 620)
128	1	5	(1, 25, −5)	(16001, 3175, 630)
140	5	1	(1, 5, −1)	(701, 135, 26)
198	−6	−1	(1, −6, 1)	(1189, 204, 35)
207	−3	−2	(1, −12, 2)	(4969, 840, 142)
210	−2	−3	(1, −18, 3)	(11341, 1908, 321)
213	−1	−6	(1, −36, 6)	(46009, 7704, 1290)
219	1	6	(1, 36, −6)	(47305, 7848, 1302)
222	2	3	(1, 18, −3)	(11989, 1980, 327)
225	3	2	(1, 12, −2)	(5401, 888, 146)
234	6	1	(1, −6, 1)	(1405, 228, 37)

7.3(e). If we let $k = u/v$, then $r = c - (u^3/v^3) = (cv^3 - u^3)/v^3$ and $s = 3k/(c - k^3)$. We have the solution

$$(x, y, z) = \left(1, \frac{3u^2v}{cv^3 - u^3}, \frac{-3uv^2}{cv^3 - u^3}\right).$$

Try different values of c and select u and v to give integer entries. This table gives some examples.

c	u	v	(x, y, z)	$(x, y, z)^{-1}$
2	5	4	(1, 100, −80)	(16001, 12700, 10080)
3	3	2	(1, −18, 12)	(649, 450, 312)
9	4	2	(1, 12, −6)	(649, 312, 150)
15	5	2	(1, −30, 12)	(5401, 2190, 888)
16	5	2	(1, 50, −20)	(16001, 6350, 2520)
19	8	3	(1, 576, −216)	(2363905, 885888, 331992)
37	10	3	(1, −900, 270)	(8991001, 2698200, 809730)
43	7	2	(1, 294, −84)	(1061929, 303114, 86520)
91	9	2	(1, −486, 108)	(4776409, 1061910, 236088)
152	16	3	(1, 288, −54)	(2363905, 442944, 82998)
166	11	2	(1, −242, 44)	(1767569, 321618, 58520)
182	17	3	(1, 2601, −459)	(217282339, 38341341, 6765660)
275	13	2	(1, 338, −52)	(4833401, 743262, 114296)

In more generality, let $v = 2$. If $u = 8w \pm 1$, then $u^3 = 8(64w^3 \pm 24w^2 + 3w) \pm 1$, so we can take $c = 64w^3 \pm 24w^2 + 3w$. If $u = 8w \pm 3$, then $u^3 = 8(64w^3 \pm 72w^2 + 27w + 3) + 3$, so we can take $c = 64w^3 \pm 72w^2 + 27w + 3$.

7.5(c). $c = 5$. $(x, y, z) = (1, -4, 2)$. The quantity $(1 - 4\theta + 2\theta^2)^5$ yields the solution $(x, y, z) = (70001, --64620, 13850)$. For $c = 25$, a solution is $(x, y, z) = (70001, 13850, -12924)$. The "inverse" solution is

$$(x, y, z) = (9375075001, 3206230550, 1096515424).$$

7.7. See also Exercise 1.8.

c	(x, y, z)	$(x, y, z)^{-1}$
7	(2, −1, 0)	(4, 2, 1)
9	(−2, 1, 0)	(4, 2, 1)
17	(18, −7, 0)	(324, 126, 49)
19	(−8, 3, 0)	(64, 24, 9)
26	(3, −1, 0)	(9, 3, 1)
28	(−3, 1, 0)	(9, 3, 1)
37	(10, −3, 0)	(100, 30, 9)
43	(−7, 2, 0)	(49, 14, 4)
91	(9, −2, 0)	(81, 18, 4)
254	(19, −3, 0)	(361, 57, 9)
614	(17, −2, 0)	(289, 34, 4)
651	(−26, 3, 0)	(676, 78, 9)
813	(28, −3, 0)	(784, 84, 9)

7.8. When $c = k^3 \pm 1$, a solution is $(x, y, z) = (k^2, k, 1)$; when $c = k^3 + 3$, a solution is $(x, y, z) = (1, k^2, -k)$, and when $c = k^3 - 3$, a solution is $(x, y, z) = (1, -k^2, k)$.

7.9(b). We find that $g_c(c+1, 3, 3) = (c+1)^3$ and $g_c(c^2+20c+1, 15c+6, 6c+15) = (c+1)^6$, so that two solutions of $g_c(x, y, z) = 1$ are

$$(x, y, z) = \left(1, \frac{3}{c+1}, \frac{3}{c+1}\right), \quad \left(1 + \frac{18c}{(c+1)^2}, \frac{3(5c+2)}{(c+1)^2}, \frac{3(2c+5)}{(c+1)^2}\right).$$

We get integer entries when $c = 2, -2$ and -4. From these, we deduce that

$$\begin{aligned} 1 = g_2(1, 1, 1) &= g_2(5, 4, 3) = g_2(1, 3, -3) \\ &= g_4(1, 1, -1) = g_4(-7, 6, -1). \end{aligned}$$

7.9(c).

$$\left(1, \frac{3}{c+1}, \frac{3}{c+1}\right)^2 = \left(1 + \frac{18c}{(c+1)^2}, \frac{3(5c+2)}{(c+1)^2}, \frac{3(2c+5)}{(c+1)^2}\right).$$

7.9(d). The transformation

$$S : (x, y, z) \longrightarrow (x + cz, x + y, y + z)$$

corresponds to multiplication of $x + y\theta + z\theta^2$ by $1 + \theta$, so if we commence the sequence with $(1, 0, 0)$, then the nth term is the triple corresponding to $(1 + \theta)^n$. $g_c(x + cz, x + y, y + z) = N(1 + \theta)N(x + y\theta + z\theta^2) = (1 + c)g_c(x, y, z)$.

It turns out that $S^2(x, y, z) = (x + cy + 2cz, 2x + y + cz, x + 2y + z)$ and $S^3(x, y, z) = ((1+c)x + 3cy + 3cz, 3x + (1+c)y + 3cz, 3x + 3y + (1+c)z)$, and also that $g_c(x + cz, y + x, z + y) = (1 + c)g_c(x, y, z) = g_c(x, y, z) + g_c(cz, x, y)$. Hence $g_c(S^3(x, y, z)) = (1 + c)^3 g(x, y, z)$. If $g_c(u, v, w) = 1$, we can divide the entries of $S^3(u, v, w)$ by $1 + c$ to get another rational solution of $g_c(x, y, z) = 1$. When $c = 2$, the solution is actually an integer. The action of S on $(1, 0, 0)$ in this case is

$$(1, 0, 0) \to (1, 1, 0) \to (1, 2, 1) \to (3, 3, 3).$$

Then we can proceed to

$$(1, 1, 1) \to (3, 2, 2) \to (7, 5, 4) \to (15, 12, 9)$$

and

$$(5, 4, 3) \to (11, 9, 7) \to (25, 20, 16) \to (57, 45, 36)$$

and so on to find infinitely many solutions.

9.3(a).

$$\begin{aligned} g(x, y, z) &= (x + y\theta_1 + z\theta_1^2)(x + y\theta_2 + z\theta_2^2)(x + y\theta_3 + z\theta_3^2) \\ &= x^3 + (\theta_1 + \theta_2 + \theta_3)x^2y + (\theta_1^2 + \theta_2^2 + \theta_3^2)x^2z \\ &\quad + (\theta_1\theta_2 + \theta_1\theta_3 + \theta_2\theta_3)xy^2 \\ &\quad + (\theta_1^2\theta_2^2 + \theta_1^2\theta_3^2 + \theta_2^2\theta_3^2)xz^2 \\ &\quad + (\theta_1\theta_2^2 + \theta_1^2\theta_2 + \theta_1\theta_3^2 + \theta_1^2\theta_3 + \theta_2\theta_3^2 + \theta_2^2\theta_3)xyz \\ &\quad + (\theta_1\theta_2\theta_3)y^3 + (\theta_1\theta_2\theta_3)(\theta_1 + \theta_2 + \theta_3)y^2z \\ &\quad + (\theta_1\theta_2\theta_3)(\theta_1\theta_2 + \theta_1\theta_3 + \theta_2\theta_3)yz^2 + (\theta_1\theta_2\theta_3)^2z^3. \end{aligned}$$

Since

$$\begin{aligned}
\theta_1 + \theta_2 + \theta_3 &= 7, \\
\theta_1\theta_2 + \theta_1\theta_3 + \theta_2\theta_3 &= 14, \\
\theta_1\theta_2\theta_3 &= 7, \\
\theta_1^2 + \theta_2^2 + \theta_3^2 &= (\theta_1 + \theta_2 + \theta_3)^2 - 2(\theta_1\theta_2 + \theta_1\theta_3 + \theta_2\theta_3) = 21, \\
\theta_1^2\theta_2^2 + \theta_1^2\theta_3^2 + \theta_2^2\theta_3^2 &= (\theta_1\theta_2 + \theta_1\theta_3 + \theta_2\theta_3)^2 - 2(\theta_1\theta_2\theta_3)(\theta_1 + \theta_2 + \theta_3) = 98, \\
\sum \theta_i\theta_j^2 &= (\theta_1 + \theta_2 + \theta_3)(\theta_1\theta_2 + \theta_1\theta_3 + \theta_2\theta_3) - 3\theta_1\theta_2\theta_3 = 77,
\end{aligned}$$

the result follows.

9.3(b). With $z = 0$, the equation becomes $x^3 + 7y^3 + 7x^2y + 14xy^2 = 1$. One solution is $(x, y, z) = (1, -1, 0)$. Taking $y = -1$ yields that

$$0 = x^3 - 7x^2 + 14x - 8 = (x - 1)(x - 2)(x - 4),$$

so we obtain $(x, y, z) = (1, -1, 0), (2, -1, 0), (4, -1, 0)$.

9.3(c). Setting $x = z = 1$ leads to

$$0 = y^3 + 9y^2 + 26y + 24 = (y + 2)(y + 3)(y + 4)$$

and the solutions $(x, y, z) = (1, -2, 1), (1, -3, 1), (1, -4, 1)$.

9.3(d). A unit in $\mathbf{Z}(\theta)$ is $1 - \theta$ and we note that

$$\begin{aligned}
&(1 - \theta) * (u + v\theta + w\theta^2) \\
&\quad = u + (v - u)\theta + (w - v)\theta^2 - w\theta^3 \\
&\quad = u + (v - u)\theta + (w - v)\theta^2 - w(7\theta^2 - 14\theta + 7) \\
&\quad = (u - 7w) + (v - u + 14w)\theta + (-v - 6w)\theta^2.
\end{aligned}$$

Thus, if $(x, y, z) = (u, v, w)$ satisfies $g(x, y, z) = 1$, then so also does

$$(x, y, z) = (u - 7w, -u + v + 14w, -v - 6w).$$

Starting with $(x, y, z) = (1, 0, 0)$, this process yields in turn

$$(x, y, z) = (1, -1, 0), (1, -2, 1), (-6, 11, -4), (22, -39, 13), \ldots.$$

Chapter 8

1.1(c). Since $t^n - 1 = (t - 1)(t^{n-1} + t^{n-2} + \cdots + t + 1)$, $\zeta^i \neq 1$, and the left side vanishes for $t = \zeta^i$, the second factor on the right must vanish for this value of t.

1.4. Use Exercise 1.2 to show that $\Pi_{i=0}^{n-1}(k - \zeta^i\theta) = k^n - c$.

1.5(a), 1.6(a). Note that $N(k - \theta) = \prod_{i=0}^{n-1}(k - \zeta^i\theta) = k^n - c$ from Exercise 1.2.

1.7(a). $N(k + \theta) = \prod_{i=0}^{n-1}(k + \zeta^i\theta) = k^n - (-\theta)^n = k^n - (-1)^n c$.

1.8.

$$N(k^2 + k\theta + \theta^2) = \prod_{i=0}^{n-1}(k^2 + k\zeta^i\theta + \zeta^{2i}\theta^2)$$
$$= \prod_{i=0}^{n-1} \frac{k^3 - \zeta^{3i}\theta^3}{k - \zeta^i\theta} = \frac{1}{k^n - c}\prod_{i=0}^{n-1}(k^3 - \zeta^{3i}\theta^3).$$

When n is not a multiple of 3, the set $\{3i : 0 \leq i \leq n-1\}$ constitutes a complete set of residues modulo n. This means that for any j between 0 and $n-1$ inclusive, there is a unique integer i in the same range for which $3i \equiv j \pmod n$. Thus $\zeta^{3i}\theta^3$ runs through all the nth roots of c^3. Thus, $\prod_{i=0}^{n-1}(k^3 - \zeta^{3i}\theta^3) = k^{3n} - c^3$. Suppose that $n = 3m$. Then ζ^{3i} is an mth root of unity. When $0 \leq i, j \leq n-1$, $\zeta^{3i} = \zeta^{3j}$ if and only if $i \equiv j \pmod m$, and so $\zeta^{3i}\theta^3$ $(0 \leq i \leq n-1)$ runs through all the mth roots of c three times. Thus $\prod_{i=0}^{n-1}(k^3 - \zeta^{3i}\theta^3) = (k^n - c)^3$. The result follows.

1.10(a).

$$\xi^2 = k^{2n-2} + 2k^{2n-3}\theta + 3k^{2n-4}\theta^2 + \cdots$$
$$+ nk^{n-1}\theta^{n-1} + (n-1)k^{n-2}\theta^n + (n-2)k^{n-3}\theta^{n+1}$$
$$+ \cdots + 2k\theta^{2n-3} + \theta^{2n-2}.$$

Now use the fact that $\theta^n = c$ to get the result.

1.10(b). When n is even and $c = k^n + 1$, then by Exercise 1.4,

$$g_c(k^{n-1}, k^{n-2}, \ldots, k, 1) = -1.$$

Using Exercise 1.7, we find that

$$g_c(k^{2n-2} + (n-1)ck^{n-2}, 2k^{2n-3} + (n-2)ck^{n-3}, \ldots, nk^{n-1}) = 1.$$

1.11. Let $\alpha = (k^n \pm 1) + k^{n-1}(\zeta^i\theta) + k^{n-2}(\zeta^i\theta)^2 + \cdots + k(\zeta^i\theta)^{n-1}$. Then

$$\alpha = k[k^{n-1} + k^{n-2}(\zeta^i\theta) + \cdots + (\zeta^i\theta)^{n-1}] \pm 1$$
$$= \frac{k(k^n - c)}{k - \zeta^i\theta} \pm 1 = \frac{\mp 2k \pm 1 \mp \zeta^i\theta}{k - \zeta^i\theta}$$
$$= \frac{\pm[(-k) - \zeta^i\theta]}{k - \zeta^i\theta}.$$

Hence

$$N(\alpha) = \frac{(-k)^n - c}{k^n - c} = 1$$

when n is even. (When n is odd, $c = k^n + 2$, this becomes $N(\alpha) = -(k^n + c)/(k^n - c) = k^n + 1$, while if n is odd, $c = k^n - 2$, it becomes $N(\alpha) = -(k^n + c)/(k^n - c) = -k^n + 1$).

1.12(a). This is a consequence of Exercise 1.10(b) with $k = 1, c = 2$. For the direct verification, we have for any positive integer r,

$$\begin{aligned}
&[r + (n-1)] + [r + (n-2)]t + \cdots + rt^{n-1} \\
&\quad = r(1 + t + \cdots + t^{n-1}) + (1 + t + \cdots + t^{n-2}) + \cdots + (1+t) + 1 \\
&\quad = \frac{1}{t-1}[r(t^n - 1) + (t^{n-1} - 1) + \cdots + (t^2 - 1) + (t-1)] \\
&\quad = \frac{1}{t-1}[r(t^n - 1) + \frac{t^n - 1}{t-1} - n] \\
&\quad = \frac{1}{(t-1)^2}[r(t^n - 1)(t-1) + (t^n - 1) - n(t-1)].
\end{aligned}$$

When $r = n$, $\theta^n = c = 2$, and $t = \zeta\theta$, this becomes $(1 - \zeta\theta)^{-2}$, so

$$N\big((2n-1) + (2n-2)\theta + \cdots + n\theta^{n-1}\big) = \prod_{i=0}^{n}(1 - \zeta^i\theta)^{-2} = (-1)^{-2} = 1.$$

1.12(b).

$$\begin{aligned}
N\big(3 + 2\theta + \cdots + 2\theta^{n-1}\big) &= \prod_{i=0}^{n-1}\left(1 + \frac{2(1-\theta^n)}{(1-\zeta^i\theta)}\right) \\
&= \prod_{i=0}^{n-1}\frac{(-1-\zeta^i\theta)}{(1-\zeta^i\theta)} \\
&= \frac{(-1)^n - 2}{1-2} = 2 + (-1)^{n-1}.
\end{aligned}$$

This assumes the value 3 when n is odd and 1 when n is even. Thus $g_2(3, 2, \ldots, 2) = 1$ when n is even.

1.14. In the solution to Exercise 1.11(a), take $r = k$ and $n = 2k - 1$ to obtain

$$\begin{aligned}
&(3k-2) + (3k-3)t + \cdots + kt^{n-1} \\
&\quad = \frac{1}{(t-1)^2}[k(t^{2k-1} - 1)(t-1) + (t^{2k-1} - 1) - (2k-1)(t-1)].
\end{aligned}$$

When $\theta^n = c = 3$ and $t = \zeta^i\theta$, this becomes

$$\frac{1}{(\zeta^i\theta - 1)^2}(2k - 2k + 1)(\zeta^i\theta - 1) + 2] = \frac{(\zeta^i\theta - 1) + 2}{(\zeta^i\theta - 1)^2} = \frac{1 + \zeta^i\theta}{(1 - \zeta^i\theta)^2},$$

and the result follows.

2.1.

$$\begin{aligned}
g_c(x, y, z, w) &= [(x + z\theta^2)^2 - \theta^2(y + w\theta^2)^2][(x - z\theta^2)^2 + \theta^2(y - w\theta^2)^2] \\
&= [(x^2 + cz^2 - 2cyw) + \theta^2(2xz - y^2 - cw^2)] \\
&\qquad \times [(x^2 + cz^2 - 2cyw) - \theta^2(2xz - y^2 - cw^2)] \\
&= (x^2 + cz^2 - 2cyw)^2 - c(2xz - y^2 - cw^2)^2.
\end{aligned}$$

2.3. Observe that

$$(x, y, z, w) * (x, -y, z, -w) = (x^2 - 2cyw + cz^2, 0, 2xz - y^2 - cw^2, 0).$$

2.4. Let $c = 1$. Then $[(x-z)^2+(y-w)^2]\cdot[(x+z)^2-(y+w)^2] = 1$. Each factor on the left must be 1, and we have that $(x, y, z, w) = (\pm 1, 0, 0, 0), (0, 0, \pm 1, 0)$. Now let $1 < c = b^4$. Then

$$(x - b^2z)^2 + b^2(y - b^2w)^2 = (x + b^2z)^2 - b^2(y + b^2w)^2 = 1.$$

The only way two integer squares can have sum or difference equal to 1 is for them to be 1 and 0. Hence $(x - b^2z)^2 = (x + b^2z)^2 = 1$ and $b^2(y - b^2w) = b^2(y + b^2w)^2 = 0$, so that $(x, y, z, w) = (\pm 1, 0, 0, 0)$.

2.6(a). $a^2w^2 = y^2 = \frac{1}{2}(y^2 + a^2w^2) = zx = z(1 + az)$.

2.6(b). The greatest common divisor of z and $1 + az$ is 1. Since their product is square, each factor is square. Since a^2 divides $z(1 + az)$ and the greatest common divisor of a and $1 + az$ is 1, a^2 must divide z. Hence there are integers r and s for which $z = a^2s^2$ and $1 + az = r^2$, whence $1 = r^2 - az = r^2 - a^3s^2$.

2.6(c). $x^2 + a^2z^2 - 2a^2yw = r^4 + a^6s^4 - 2a^3r^2s^2 = (r^2 - a^3s^2)^2 = 1$ and $y^2 + a^2w^2 = 2a^2r^2s^2 = 2xz$.

2.6(d). $r^2 - a^3s^2 = 1$ is satisfied by

$$(a; r, s) = (2; 3, 1), (3; 26, 5), (5; 930249, 83204).$$

[For the last, we look for a solution of $x^2 - 5y^2 = 1$ where y is divisible by 5; we find that $930249 + 5 \times 83204\sqrt{5} = (9 + 4\sqrt{5})^5$.] These yield

$$\begin{aligned} &g_4(9, 6, 4, 3) \\ &\quad = g_9(676, 390, 225, 130) \\ &\quad = g_{25}(865363202001, 387002188980, 173072640400, 77400437796) \\ &\quad = 1. \end{aligned}$$

2.7. We have $a^2w^2 = y^2 = zx = z(az - 1)$. Hence $z = a^2s^2$ and $az - 1 = r^2$ for some integers r and s, whence $r^2 - a^3s^2 = -1$. On the other hand, if $r^2 - a^3s^2 = -1$ and $(x, y, z, w) = (r^2, ars, a^2s^2, rs)$, then $x^2 + a^2z^2 - 2a^2yw = (r^2 - a^3s^2)^2 = 1$ and $y^2 + a^2w^2 = 2xz$, so we get the desired solution. Let $a = 5$. $x^2 - 5y^2 = -1$ is satisfied by $(x, y) = (2, 1)$ and $(2 + \sqrt{5})^5 = (682, 305)$. We find that $r^2 - 125s^2 = -1$ is satisfied by $(r, s) = (682, 61)$. This leads to $g_{25}(465124, 208010, 93025, 41602) = 1$.

2.9. $g_8(3, 2, 1, 0.5) = 1$.

2.10(a).

(x, y, z, w)	Type	$g_2(x, y, z, w)$
(1, 1, 1, 1)*	$(-, -, -)$	-1
(2, 1, 1, 1)	$(+, -, -)$	2
(2, 2, 1, 1)	$(-, +, -)$	-4
(2, 2, 2, 1)	$(-, -, +)$	8
(3, 3, 2, 1)	$(-, +, +)$	23
(3, 2, 2, 1)	$(+, -, +)$	9
(3, 2, 1, 1)	$(+, +, -)$	9
(4, 3, 2, 1)	$(+, +, +)$	94
(5, 4, 3, 2)	$(+, +, +)$	49
(6, 5, 4, 3)	$(+, +, +)$	14
(7, 6, 5, 4)*	$(-, +, +)$	1
(9, 7, 6, 5)	$(+, -, +)$	7
(11, 9, 7, 6)	$(+, +, -)$	7
(13, 11, 9, 7)	$(-, +, +)$	79
(15, 12, 10, 8)	$(+, +, +)$	113
(16, 13, 11, 9)	$(+, -, +)$	18
(18, 15, 12, 10)	$(+, +, +)$	46
(19, 16, 13, 11)	$(-, +, -)$	-7
(35, 29, 42, 20)	$(+, +, +)$	207
(36, 30, 25, 21)	$(+, +, +)$	28
(37, 31, 26, 22)*	$(+, +, -)$	-1

2.10(b)

(x, y, z, w)	Type	$g_3(x, y, z, w)$
(1, 1, 1, 1)	$(-, -, -)$	-8
(2, 1, 1, 1)*	$(+, -, -)$	1
(2, 2, 1, 1)	$(-, +, -)$	-2
(2, 2, 2, 1)	$(-, -, +)$	13
(3, 2, 1, 1)	$(+, +, -)$	-3
(3, 2, 2, 1)	$(+, -, +)$	6
(3, 3, 2, 1)	$(-, +, +)$	9
(4, 3, 2, 1)	$(+, +, +)$	52
(5, 4, 3, 2)	$(-, +, +)$	4
(7, 5, 4, 3)*	$(+, -, +)$	1
(9, 7, 5, 4)	$(-, +, -)$	-3
(16, 12, 9, 7)	$(+, +, -)$	-2
(18, 14, 11, 8)	$(-, -, +)$	33
(34, 26, 20, 15)	$(-, -, +)$	13
(50, 38, 29, 22)*	$(-, -, +)$	1

2.11. $p_c(u, 0, v, 0) = u^2 + cv^2$ and $q_c(u, 0, v, 0) = 2uv$.

2.12(b).

$$q_c(u, r, v, s) = 0 \Leftrightarrow 0 = 2uv - r^2 - (cv^4/r^2) \Leftrightarrow$$
$$0 = (r^2 - uv)^2 - v^2(u^2 - cv^2) = (r^2 - uv)^2 - v^2$$
$$= (r^2 - uv - v)(r^2 - uv + v).$$

2.12(c). Solutions with $r^2 = (u + 1)v$:

$$(c; u, r, v, s) = (5; 161, 108, 72, 48), \left(8; 3, 2, 1, \tfrac{1}{2}\right), (12; 7, 4, 2, 1),$$
$$(14; 15, 8, 4, 2), (39; 1249, 500, 200, 80).$$

Solutions with $r^2 = (u - 1)v$:

$$(c; u, r, v, s) = (2; 3, 2, 2, 2), (3; 2, 1, 1, 1), (18; 17, 8, 4, 2),$$
$$(20; 9, 4, 2, 1), \left(24; 5, 2, 1, \tfrac{1}{2}\right).$$

2.13. $g_c(8k^4 + 1, 4k^3, 2k^2, k) = 1$ (Type A), when $c = 16k^4 - 4$.

2.14(b). Let $c \equiv 2 \pmod 8$. Since y must be even, $x^2 + 2z^2 \equiv 7 \pmod 8$, which is impossible. If $c \equiv 6 \pmod 8$, then $c = 8k + 6 = 2(4k + 3)$ is divisible by a prime congruent to 3 (mod 4). Since x^2 cannot be congruent to -1 for such a prime, the desired result follows.

2.15(a). If $p < 0$, then $q\sqrt{c} < p < 0$, so that $p^2 < cq^2 = p^2 - 1$, giving a contradiction.

2.15(b). The fundamental quadratic solution for $c = 3$ is (2, 1) and this cannot be had since $p \equiv x^2 \not\equiv 2 \pmod 3$. Similarly, the fundamental quadratic solutions for $c = 6$ and for $c = 8$ are, respectively, (5, 2) and (3, 1). Neither 5 nor 3 is congruent to a square modulo the corresponding value of c.

2.16. Let $r^2 - cs^2 = -1$ and suppose $(x, y, z, w) = (r, 0, s, 0)$.

2.18(a).

$$u^4 - \left(d^2 + \frac{2d}{v^2}\right)v^4 = u^4 - (dv^2+1)^2 + 1 = (u^2 - dv^2 - 1)(u^2 + dv^2 + 1) + 1 = 1.$$

2.18(b). If $v = 1$, then we can take $d = \pm u^2 - 1$, so that $x^4 - cy^4 = 1$ with $c = u^4 - 1$ has solution $(x, y) = (u, 1)$.

u	v	d	$c = d^2 + 2d/v^2$	$u^4 - cv^4 = 1$
3	2	2	5	$3^4 - 5 \cdot 2^4 = 1$
5	2	6	39	$5^4 - 39 \cdot 2^4 = 1$
7	2	12	150	$7^4 - 150 \cdot 2^4 = 1$
9	2	20	410	$9^4 - 410 \cdot 2^4 = 1$
80	3	711	505679	$80^4 - 505679 \cdot 3^4 = 1$
82	3	747	558175	$82^4 - 558175 \cdot 3^4 = 1$
63	4	248	61535	$63^4 - 61535 \cdot 4^4 = 1$
65	4	264	69729	$65^4 - 69729 \cdot 4^4 = 1$
182	5	−1325	1755519	$182^4 - 1755519 \cdot 5^4 = 1$
624	5	15575	242581871	$624^4 - 242581871 \cdot 5^4 = 1$
626	5	15675	245706879	$626^4 - 245706879 \cdot 5^4 = 1$

3.1. The units of $\mathbf{Q}(\sqrt[4]{-1})$ are θ^i, where $\theta^4 = -1$ and $0 \le i \le 7$. These units give rise to the solutions $(\pm 1, 0, 0, 0)$, $(0, \pm 1, 0, 0)$, $(0, 0, \pm 1, 0)$, $(0, 0, 0, \pm 1)$. But there are other solutions, such as $(1, 1, 0, -1)$ and $(3, 2, 0, -2) = (1, 1, 0, -1)^2$. Any solution (x, y, z, w) gives rise to related solutions $(y, x, -w, -z)$ and $(z, w, -x, -y)$.

3.4. $x = 1$ leads to $z^2 = 2yw$ and $2z = y^2 - 3w^2$; some experimentation yields $(x, y, z, w) = (1, 8, 8, 4)$. On the other hand, $w = 0$ leads to $x^2 - 3z^2 = 1$, $y^2 = 2xz$ and the solution $(x, y, z, w) = (2, 2, 1, 0)$.

3.5. We require $w = z^2/2y$, $d = (y^2 - 2z)4y^2/z^4$. For example, let $(d; x, y, z, w) = (4t(t^3 - 1); 1, 2t^2, 2t, 1)$.

3.6. $(d; x, y, z, w) = (3; 2, 2, 1, 0), (20; 9, 6, 2, 0)$.

3.7. The greatest common divisor of d and x must be 1, so that $z = dr$ for some value of r. Hence $x^2 - d^3r^2 = \pm 1$ and $w^2 = -2xr$.

4.2. Note that, when $\theta^5 = c$,

$$x + y\theta^2 + z\theta^4 + u\theta^6 + v\theta^8 = x + cu\theta + y\theta^2 + cv\theta^3 + z\theta^4,$$
$$x + y\theta^3 + z\theta^6 + u\theta^9 + v\theta^{12} = x + cz\theta + c^2v\theta^2 + y\theta^3 + cu\theta^4,$$
$$x + y\theta^4 + z\theta^8 + u\theta^{12} + v\theta^{16} = x + c^3v\theta + c^2u\theta^2 + cz\theta^3 + y\theta^4.$$

For example, some solutions to $g_2(x, y, z, u, v) = 1$ are

$$\begin{aligned}(x, y, z, u, v) &= (1, 1, 1, 1, 1) * (1, 1, 1, 1, 1)\\ &= (9, 8, 7, 6, 5) = (9, 2 \times 4, 7, 2 \times 3, 5)\end{aligned}$$

and

$$\begin{aligned}(x, y, z, w) &= (1, 1, 0, 1, 0) * (1, 1, 0, 1, 0)\\ &= (1, 4, 1, 2, 2) = (1, 2 \times 2, 1, 2 \times 1, 2),\end{aligned}$$

whence we find that $g_4(9, 7, 5, 4, 3) = g_4(1, 1, 2, 2, 1) = 1$.

4.3.

$$\begin{aligned}c = 2 &: (1, 1, 0, 1, 0), (1, 1, 1, 1, 1), (1, -2, 1, 0, 0), (1, 4, 1, 2, 2),\\ &\quad (5, 4, 4, 3, 3), (9, 8, 7, 6, 5), (33, 29, 25, 22, 19).\\ c = 3 &: (1, 1, 0, 1, 0), (7, 6, 5, 4, 3), (1, 5, 1, 2, 2).\\ c = 4 &: (1, 1, 0, -1, 0), (9, 7, 5, 4, 3), (1, 1, 2, 2, 1).\\ c = 5 &: (1, 0, 0, 1, -1), (76, 55, 40, 29, 21).\end{aligned}$$

5.3(b). $g_{13867245}(31, 2, 0, 0, 0, 0) = g_{20179187}(33, 2, 0, 0, 0, 0) = 1.$

Comments on the Explorations

Exploration 8.1. Some experimentation yields

$$g_{80}(161, 54, 18, 6) = g_{82}(163, 54, 18, 6) = 1$$

and

$$g_{255}(511, 128, 32, 8) = g_{257}(513, 128, 32, 8) = 1.$$

This suggests the generalization, for $c = k^4 \pm 1$:

$$g_c(2k^4 \pm 1, 2k^3, 2k^2, 2k) = 1.$$

Can this be generalized further, either for the quartic case or with respect to degree?

Exploration 8.2. For small values of c, we present the smallest Type A solution found along with a Type B solution corresponding to the fundamental solution of the corresponding quadratic Pell's equation.

c	(x, y, z, w) A	(x, y, z, w) B
2	(3, 2, 2, 2)	(1, 0, 1, 0)
3	(2, 1, 1, 1)	(7, 5, 4, 3)
4	(9, 6, 4, 3)	—
5	(6, 4, 3, 2)	(2, 0, 1, 0), (3, 2, 0, 0)
6	(53, 34, 22, 14)	(5, 0, 2, 0)
7	(43, 26, 16, 10)	(13, 8, 5, 3), (1, 1, −1, 0)
8	(33, 20, 12, 7)	(3, 0, 1, 0)
9	(676, 390, 225, 130)	—
10		(3, 0, 1, 0)
11		(10, 0, 3, 0)
12	(7, 4, 2, 1)	(7, 0, 2, 0)
13		(18, 0, 5, 0)
14	(15, 8, 4, 2)	(15, 0, 4, 0)
15	(31, 16, 8, 4)	(8, 4, 2, 1), (2, 1, 0, 0)
17	(33, 16, 8, 4)	(4, 0, 1, 0), (2, 1, 0, 0), (8, 4, 2, 1)

One strategy is to use the equation $1 = (x - z\sqrt{c})^2 + \sqrt{c}(y - w\sqrt{c})^2$ to deduce that $|x - z\sqrt{c}| < 1$ and $|y - w\sqrt{c}| < c^{-1/4}$ along with parity considerations to reduce the number of quadruples (x, y, z, w) to be tried.

Exploration 8.3. One way to generalize is to let (u, v) be a fundamental solution of $u^2 - cv^2 = \pm 1$, and let

$$x_n + z_n\sqrt{c} = (h + k\sqrt{c})(u + v\sqrt{c})^n,$$
$$y_n + w_n\sqrt{c} = (r + s\sqrt{c})(u + v\sqrt{c})^n.$$

With $p_n = x_n^2 + cz_n^2 - 2cy_nw_n$ and $q_n = 2x_nz_n - y_n^2 - cz_n^2$, we have $p_n + q_n\sqrt{c} = (u + v\sqrt{c})^{2n}[(h^2 + ck^2 - 2crs) - \sqrt{c}(2hk - r^2 - cs^2)]$. Now select (h, r, k, s) such that $g_c(h, r, k, s) = 1$. The examples given correspond to $c = 2$, $(u, v) = (1, 1)$, $(h, r, k, s) = (1, -2, 0, 2)$ with $x_n + \sqrt{2}z_n = (1 + \sqrt{2})^n$ and $y_n + \sqrt{2}w_n = -2(1 - \sqrt{2})(1 + \sqrt{2})^n = 2(1 + \sqrt{2})^{n-1}$, as well as to $c = 3$, $(u, v) = (2, 1)$, $(h, r, k, s) = (1, -1, 0, 1)$ with $x_n + \sqrt{3}z_n = (2 + \sqrt{3})^n$ and $y_n + \sqrt{3}w_n = (-1 + \sqrt{3})(2 + \sqrt{3})^n$.

Exploration 8.4. We have $g_{-4}(6, 4, 1, -1) = 0$. However, when c is a positive nonsquare, then $x^2 + cz^2 - 2cyw = 2xz - y^2 - cw^2 = 0$ leads to $(x + \sqrt{c}z)^2 = \sqrt{c}(y + \sqrt{c}w)^2$. Since x, y, z, w are rational and $\sqrt{c}$ is irrational, we have that

$$\theta = \frac{x + \sqrt{c}z}{y + \sqrt{c}w}.$$

This can be rewritten as $\sqrt{c}(w\theta - z) = x - y\theta$, squared and transformed to a quadratic equation in θ with integer coefficients. However, the polynomial equation of lowest degree with integer coefficients with root θ has fourth degree.

Exploration 8.7. Cuong Nguyen, while an undergraduate at the University of Toronto, found the following polynomial solutions to $p^2 - cq^2 = \pm 1$, where $p = x^2 + cz^2 - 2cyw$ and $q = 2xz - y^2 - cw^2$:

c	(x, y, z, w)	(p, q)
$t^4 \pm 1$	$(t^2, 0, 1, 0)$	$(2t^4 \pm 1, 2t^2)$
	$(t^3, t^2, t, 1)$	$(\mp t^2, \mp 1)$
	$(t^3, t^2, -t, -1)$	$(4t^6 \pm 3t^2, -4t^4 \mp 1)$
	$(1, -2t^3, 0, 2t)$	$(8t^8 \pm 8t^4 + 1, -8t^6 \mp 4t^2)$
$t^4 \pm 2t$	$(t^3 \pm 1, t^2, t, 1)$	(1,0)
$t^4 \pm 2$	$(t^4 \pm 1, t^3, t^2, t)$	(1,0)
	$(1, t^3, 0, -t)$	$(2t^8 \pm 4t^4 + 1, -2t^6 \mp 2t^2)$
	$(t^4 \pm 1, 0, t^2, 0)$	$(2t^8 \pm 4t^4 + 1, 2t^6 \pm 2t^2)$
$t^4 \pm t$	$(1 \pm 2t^3, 0, \pm 2t, 0)$	$(8t^6 \pm 8t^3 + 1, 8t^4 \pm 4t)$
	$(1, -2t^2, 0, 2)$	$(8t^6 \pm 8t^3 + 1, -8t^4 \mp 4t)$
$s^4t^4 - s$	$(1, -2s^2t^3, 0, 2t)$	$(8s^6t^8 - 8s^3t^4 + 1, -8s^4t^6 + 4st^2)$
	$(2s^3t^4 - 1, 0, 2st^2, 0)$	$(8s^6t^8 - 8s^3t^4 + 1, 8s^4t^6 - 4st^2)$
$s^4t^4 - 2s$	$(1, -s^2t^3, 0, t)$	$(2s^6t^8 - 4s^3t^4 + 1, -2s^4t^6 + 2st^2)$
	$(s^3t^4 - 1, 0, st^2, 0)$	$(2s^6t^8 - 4s^3t^4 + 1, 2s^4t^6 - 2st^2)$

Exploration 8.8.

(x, y, z, u, v, w)	(ρ, σ)	(ξ, η, ζ)	$g_2(x, y, z, u, v, w)$
(1, 1, 0, 0, 0, 0)	(1, 1)	(1, −1, 0)	−1
(1, 0, 0, 1, 0, 0)	(7, 5)	(−1, 0, 0)	−1
(1, 0, 1, 0, 1, 0)	(1, 0)	(5, 4, 3)	+1
(1, 1, 1, 0, 0, 0)	(3, −2)	(1, 1, 1)	+1
(1, 1, 1, 1, 1, 1)	(1, 1)	(−1, −1, −1)	−1
(3, 2, 2, 2, 2, 2)	(3, 2)	(1, 0, 0)	+1
(11, 10, 9, 8, 7, 6)	(3, 2)	(5, 4, 3)	+1
(145, 138, 126, 108, 90, 78)	(1, 0)	(1, 0, 0)	+1

The last solution is obtained by *-multiplication. Let X, Y, Z be the respective solutions (3, 2, 2, 2, 2, 2), (1, 1, 1, 0, 0, 0), (1, 0, 1, 0, 1, 0). By looking at the solutions to the induced quadratic and cubic Pell's equations, we find that $X^2Y^2Z^{-1} =$ (145, 138, 126, 108, 90, 78) induces the trivial solution to the lower-degree equations. Is this the solution with the smallest positive integers?

Chapter 9

1.1. The multiplication table, modulo 11, is

	0	1	2	3	4	5	6	7	8	9	10
0	0	0	0	0	0	0	0	0	0	0	0
1	0	1	2	3	4	5	6	7	8	9	10
2	0	2	4	6	8	10	1	3	5	7	9
3	0	3	6	9	1	4	7	10	2	5	8
4	0	4	8	1	5	9	2	6	10	3	7
5	0	5	10	4	9	3	8	2	7	1	6
6	0	6	1	7	2	8	3	9	4	10	5
7	0	7	3	10	6	2	9	5	1	8	4
8	0	8	5	2	10	7	4	1	9	6	3
9	0	9	7	5	3	1	10	8	6	4	2
10	0	10	9	8	7	6	5	4	3	2	1

1.3. The solutions of $x^2 - y^2 \equiv 1$ are (2, 5), (7, 9), (4, 9), (9, 5), (10, 0), (9, 6), (4, 2), (7, 2), (2, 6), (1, 0).

The solutions of $x^2 - 3y^2 \equiv 1$ are (2, 1), (7, 4), (4, 4), (9, 1), (10, 0), (9, 10), (4, 7), (7, 7), (2, 10), (1, 0).

The solutions of $x^2 - 4y^2 \equiv 1$ are (2, 3), (7, 1), (4, 1), (9, 3), (10, 0), (9, 8), (4, 10), (7, 10), (2, 8), (1, 0).

1.3, 1.6, 1.7. The solutions of $x^2 - 2y^2 \equiv 1$ are (3, 2), (6, 1), (0, 4), (5, 1), (8, 2), (10, 0), (8, 9), (5, 10), (0, 7), (6, 10), (3, 9), (1, 0), listed in ascending $*$-powers of (3, 2), so that, for example, $(8, 2) \equiv (3, 2)^5$.

2.2(c) Suppose that $ai \equiv aj$ with $0 \leq i < j \leq p-1$. Then $a(i-j) \equiv 0$, so that $i \equiv j$. (Since p divides the product of a and $i - j$, it divides one of the factors.) Thus, $0, a, 2a, \ldots, (p-1)a$ are p incongruent elements. Since a complete set of incongruent elements has exactly p entries, the result follows.

2.3. Since $2x \equiv u + v$ and $2ry \equiv v - u$, and $2 \not\equiv 0 \pmod{p}$, we can solve these congruences uniquely for x and y by Exercise 2.2. Thus, given any solution to $x^2 - r^2y^2 \equiv 1$, we can find a pair of reciprocals and vice versa. Since each nonzero element of $\mathbf{Z}_p$ has a reciprocal, there are $p - 1$ such ordered pairs, and the result follows.

2.4(a).

$$(x_1, y_1) * (x_2, y_2) = (x_1x_2 + dy_1y_2, x_1y_2 + x_2y_1)$$

corresponds to

$$(x_1x_2 + r^2y_1y_2) - r(x_1, y_2 + x_2y_1) = (x_1 - ry_1)(x_2 - ry_2).$$

2.5(a). a^{-1} and b^{-1} satisfy $aa^{-1} = a^{-1}a \equiv 1$ and $bb^{-1} = b^{-1}b \equiv 1 \pmod{p}$. Therefore,

$$(ab)(b^{-1}a^{-1}) = a(bb^{-1})a^{-1} \equiv a \cdot 1 \cdot a^{-1} = aa^{-1} \equiv 1$$

modulo p, so $b^{-1}a^{-1}$ is the inverse of ab.

2.5(b).

$$\begin{aligned} x^2 - dy^2 \equiv 1 &\iff dy^2 \equiv x^2 - 1 = (x-1)(x+1) \\ &\iff d \equiv y^{-1}(x-1) \cdot y^{-1}(x+1) \\ &\qquad = (xy^{-1} - y^{-1})(xy^{-1} + y^{-1}). \end{aligned}$$

3.2(d). Note that we can write $f(t) \equiv (t - a_1)g(t)$ for some polynomial $g(t)$ over $\mathbf{Z}_p$. Then $0 \equiv f(a_2) \equiv (a_2 - a_1)g(a_2)$. Since $a_2 - a_1 \not\equiv 0$, $g(a_2) \equiv 0$, so that $g(t)$ is divisible by $t - a_2$ modulo p.

3.4. The little Fermat theorem says that each nonzero element of $\mathbf{Z}_p$ is a root of the polynomial $t^{p-1} - 1$. Now use the factor theorem.

3.5(c). For each prime $p = 11, 13$, k turns out to be a divisor of $p - 1$.

3.6. First, we need to establish that such an integer k exits. Since $\mathbf{Z}_p$ is finite, the powers a^n with integer n cannot all be distinct. Select r and s with $r < s$ and $a^r \equiv a^s$. Then $a^{s-r} \equiv 1$, so that some positive power of a is congruent to 1. There is a smallest positive exponent k for which $a^k \equiv 1$. Suppose that $p - 1 = uk + v$, where u and v are nonnegative integers with $0 \le v < k$. Then

$$1 \equiv a^{p-1} \equiv a^{uk+v} \equiv (a^k)^u a^v \equiv a^v.$$

Since k is the smallest positive integer with $a^k \equiv 1$, we must have $v = 0$.

3.7(b). a and p^r are relatively prime if and only if a is not divisible by p. For $0 \le a \le p^r - 1$, this occurs only if a is not one of the p^{r-1} multiples of p.

3.7(d). $\phi(1) = \phi(2) = 1$. Suppose that $m \ge 3$. Then when m is even, $\frac{1}{2}m$ and m are not relatively prime. When $1 \le a \le m - 1$, a and m are relatively prime if and only if $m - a$ and m are relatively prime; when this occurs, a and $m - a$ are unequal. Thus, the relatively prime positive integers not exceeding m come in pairs of distinct elements, so that $\phi(m)$ is even. Thus, $\phi(m)$ is odd if and only if $m = 1, 2$.

3.9. Let Q_k be the set of positive integers x not exceeding m for which the greatest common divisor of x and m is k. Each of the numbers $1, 2, \ldots, m$ belongs to exactly one Q_k, with k a divisor of m. If $x \in Q_k$, then $1 \le x/k \le m/k$, and the greatest common divisor of x/k and m/k is 1. On the other hand, suppose that $1 \le y \le m/k$ and y and m/k are relatively prime. Then $1 \le ky \le m$, and the greatest common divisor of ky and m is k. It follows from this pairing that $\#Q_k = \phi(m/k)$, from which the result follows.

3.10(a). Let $p - 1 = kl$. Then $t^{p-1} - 1 = t^{kl} - 1 = (t^k - 1)\big(t^{k(l-1)} + t^{k(l-2)} + \cdots + t^k + 1\big)$.

3.10(b). We know that $t^{p-1} - 1$ has exactly $p - 1$ roots; each one is a root of either $t^k - 1$ or $q(t)$. Since $t^k - 1$ and $q(t)$ can have no more roots that their respective degrees k and $p - 1 - k$, and since $p - 1 = k + (p - 1 - k)$, $t^k - 1$ must have exactly k roots and $q(t)$ exactly $p - 1 - k$ roots.

3.11. Suppose that $a^k \equiv 1$, while $a^i \not\equiv 1$ for $1 \le i \le k-1$. Then $(a^i)^k \equiv (a^k)^i \equiv 1$, so that $\{1, a, a^2, \ldots, a^{k-1}\}$ are k distinct roots of $t^k - 1$.

Let $1 \le i \le k$ and suppose that r is the greatest common divisor of k and i; let $k = rs$ and $i = rj$. Then $(a^i)^s \equiv a^{jrs} \equiv a^{kj} \equiv 1$. If $r > 1$, then $s < k$, so that a^i does not belong to the exponent k. On the other hand, suppose $r = 1$. Let a^i belong to the exponent w. Then $a^{iw} \equiv 1$, so that by the reasoning of Exercise 3.6, k must be a divisor of iw. Since k and i are relatively prime and $w \le k$, it follows that $w = k$. We obtain the desired result.

3.13(b). Primes and primitive roots: $(3 : 2)$, $(5 : 2, 3)$, $(7 : 3, 5)$, $(11 : 2, 6, 7, 8)$, $(13 : 2, 6, 7, 11)$.

3.14. By Exercise 2.4, there is a one-to-one correspondence between solutions of $x^2 - r^2y^2 \equiv 1$ and nonzero integers w. Suppose g is a primitive root modulo p and $(u, v) \sim g$. Then $(u, v)^i \sim g^i$ yield the $p-1$ incongruent solutions of Pell's congruence.

3.15(b). Suppose the result holds for $1 \le n \le m$. Then

$$\begin{aligned}
(x_{m+1}, y_{m+1}) &\equiv (x_1, y_1) * (x_m, y_m)\\
&= (x_1x_m + r^2y_1y_m, x_1y_m + x_my_1)\\
&= \Big(2^{-2}\big(g^{m+1} + g^{m-1} + g^{-(m-1)} + g^{-(m+1)}\big)\\
&\qquad + r^2(2^{-2}r^{-2})\big(g^{m+1} - g^{m-1} - g^{-(m-1)} + g^{-(m+1)}\big),\\
&\qquad 2^{-2}r^{-1}\big(g^{m+1} + g^{m-1} - g^{-(m-1)} - g^{-(m+1)}\\
&\qquad + g^{m+1} - g^{m-1} + g^{-(m-1)} - g^{-(m+1)}\big)\Big)\\
&= \big(2^{-2}(2g^{m+1} + 2g^{-(m+1)}), 2^{-2}r^{-1}(2g^{m+1} - 2g^{-(m+1)})\big)\\
&= \Big(2^{-1}\big(g^{m+1} + g^{-(m+1)}\big), (2r)^{-1}\big(g^{m+1} - g^{-(m+1)}\big)\Big)
\end{aligned}$$

4.1. See Exercise 3.4.11.

4.3(b). If r and s are selected with $1 \le r < s$ and $(u, v)^r \equiv (u, v)^s$, then $*$-multiplying both sides of the equation by $(u, -v)^r$ yields $(1, 0) \equiv (u, v)^{s-r}$.

4.3(c). If $x^2 - dy^2 \equiv 1$ and $x \equiv 1$ or $x \equiv p - 1$, then y must be congruent to 0. Hence there is a unique solution with $x \equiv 1$ and with $x \equiv p - 1$. Suppose $x^2 \not\equiv 1$. Then $y^2 \equiv d^{-1}(x^2 - 1)$ is a quadratic equation in y that has two distinct solutions. Now $(u, v)^{m-1} \equiv (u, v)^m * (u, -v) \equiv (1, 0) * (u, -v) \equiv (u, -v)$. By induction, it can be shown that $(u, v)^{m-i} \equiv (u, -v)^i$, and the result follows from this.

4.4(b). From Exercise 3.4.3 we see that $U_k(t)$ is a polynomial of degree $k - 1$ with leading coefficient 2^{k-1} that is not congruent to 0.

4.4(c). $(u_i, v_i)^{2k} \equiv (u, v)^{i(2k)} \equiv (u, v)^{2ki} \equiv (1, 0)^i \equiv (1, 0)$, whence $T_{2k}(u_i) \equiv 1$. Hence $(u_i - 1)(u_i + 1)U_k(u_i) \equiv 0$. Among the $2k$ values of u_i, at most two are equal to ± 1. The remaining $2(k-1)$ are roots of $U_k(t)$; since each u_i appears only twice, all roots of $U_k(t)$ must be involved.

4.4(d). Suppose (u', v') belongs to the exponent $2k$. Then $U_k(u') = 0$, so that by (c), $u' = u_i$ for some i. Then $v' = \pm v_i$, so that (u', v') is equal to one of (u_i, v_i) or $(u_i, -v_i) = (u_{2k-i}, v_{2k-i})$. In a way analogous to that of Exercise 3.11, it can be shown that (u_i, v_i) belongs to the exponent $2k$ if and only if $\gcd(i, 2k) = 1$.

4.7(b). The elements of each S_i are distinct, since $(a, b) * (u, v)^i \equiv (a, b) * (u, v)^j$ implies that

$$(u, v)^i \equiv (a, -b) * (a, b) * (u, v)^i \equiv (a, -b) * (a, b) * (u, v)^j \equiv (u, v)^j.$$

4.7(c). Suppose that $S_r \cap S_s$ is nonvoid. Then there are pairs $(a_r, b_r) \in S_r$ and $(a_s, b_s) \in S_s$ (used as (a, b) in the definition of these sets) and indices j and k with $j < k$ and

$$(a_r, b_r) * (u, v)^j \equiv (a_s, b_s) * (u, v)^k,$$

so that

$$(a_r, b_r) \equiv (a_s, b_s) * (u, v)^{k-j}$$

and

$$(a_s, b_s) \equiv (a_r, b_r) * (u, v)^{m+j-k}.$$

This says that $(a_r, b_r) \in S_s$ and $(a_s, b_s) \in S_r$, contradicting the choice of either (a_r, b_r) or (a_s, b_s).

5.1(a). $G(3, 2) = \{(1, 0), (2, 0), (0, 1), (0, 2)\}$.

5.1(c). $x^2 - 2y^2 \equiv 1 \pmod 3$ is satisfied by $(x, y) \equiv (a, 0)$ and $(0, a)$, where $a \equiv 1, 2, 4, 5, 7, 8 \pmod 9$.

$G(9, 5) = \{(1, 0), (1, 3), (1, 6), (8, 0), (8, 3), (8, 6), (0, 4), (3, 4), (6, 4), (0, 5), (3, 5), (6, 5)\}$.

5.1(d). $G(9, 2) = \{(1, 0), (1, 3), (1, 6), (8, 0), (8, 3), (8, 6), (0, 7), (3, 7), (6, 7), (0, 2), (3, 2), (6, 2)\}$.

5.2(a). Since $2(a-1) \geq a$, $p^{2(a-1)} \equiv 0 \pmod{p^a}$. Thus

$$\begin{aligned} w^2 - dz^2 \equiv 1 \pmod{p^a} &\Longleftrightarrow u^2 + 2usp^{a-1} - dv^2 - 2dvtp^{a-1} \equiv 1 \pmod{p^a} \\ &\Longleftrightarrow 2us - 2dvt + c \equiv 0 \pmod p. \end{aligned}$$

5.2(b). Since $u^2 - dv^2 \equiv 1 \pmod{p^{a-1}}$, at least one of u and v is not divisible by p. If u is not divisible by p, then for each of the p choices of t, we can solve the congruence $2us \equiv 2dvt - c$ for a unique value of s. Similarly, if v is not divisible by p, then for each of the p choices of s, we can solve the congruence for a unique value of t. Thus, for each (u, v), there are p choices of the (s, t) for which $w^2 - dz^2 \equiv 1 \pmod{p^a}$.

Glossary

a **is congruent to** b **modulo** m (Symbolically: $a \equiv b \pmod{m}$): $a - b$ is a multiple of m

$[a, b]$: The closed interval whose endpoints are the real numbers a and b, namely $\{x : a \le x \le b\}$.

$\#S$: The number of elements in the set S

$|XY|$: Length of the line segment XY

$\mathbf{Q}(\alpha)$: The set of numbers of the form $p(\alpha)$, where α is a real number and p is a polynomial with rational coefficients.

$\mathbf{Z}_p$: For a prime p, $\mathbf{Z}_p = \{0, 1, 2, \ldots, p - 1\}$ is a field in which the arithmetic operations are defined modulo p.

algebraic number/integer: An algebraic number is a root of a polynomial whose coefficients are integers; it is an algebraic integer if the leading coefficient of the polynomials is 1.

ceiling: The ceiling of the real number x, denoted by $\lceil x \rceil$ is that integer n for which $n - 1 < x \le n$.

closed: A set of numbers is closed under addition (resp. multiplication) is it contains along with any two elements their sum (resp. product).

common divisor: A common divisor of two numbers is any number for which each of the two numbers is an integer multiple. The greatest common divisor is the largest of such numbers. The greatest common divisor of a and b is denoted by $\gcd(a, b)$.

common fraction: A common fraction is a rational number written with an integer numerator divided by an integer denominator.

complete set of residues modulo m: $\{r_1, r_2, \ldots, r_m\}$ is a complete set of residues if, for each integer n, there is exactly one index i for which $n \equiv r_i \pmod{m}$.

coprime: Two integers are coprime if their greatest common divisor is equal to 1.

cubic polynomial: A cubic polynomial is one for which the term of maximal degree has degree 3.

de Moivre's theorem; If n is an integer and i is a square root of -1, then $(\cos\theta + i\sin\theta)^n = \cos n\theta + i\sin n\theta$.

diophantine equation: A diophantine equation is an algebraic equation with integer coefficients for which integer solutions are sought.

divides: " a divides b" (written $a|b$) means that a is a divisor of b or that b is a multiple of a.

discriminant: The discriminant of a quadratic polynomial $ax^2 + bx + c$ is the quantity $b^2 - 4ac$.

Factor theorem: Let $p(x)$ be a polynomial in a single variable x with coefficients in a field (usually real or complex). Then $p(r) = 0$ if and only if $p(x) = (x - r)q(x)$ for some polynomial $q(x)$.

field: A field is a set of entities upon which there are two operations defined, called addition and multiplication. Both operations are commutative and associative; multiplication is distributive over addition (i.e., $a(b + c) = ab + ac$ for any three elements); each element has an additive inverse (or opposite) and each nonzero element has a multiplicative inverse (or reciprocal). A quadratic field $\mathbf{Q}(\sqrt{d})$ is the set of numbers of the form $a + b\sqrt{d}$, where a and b are rationals.

floor: The floor of the real number x, denoted by $\lfloor x \rfloor$ is that integer n for which $n \le x < n + 1$.

fractional part: The fractional part of a real number is the number obtained by subtracting its floor from it.

homomorphism: A homomorphism ϕ from one field or ring to another is a function that preserves sums and products in the sense that $\phi(a+b) = \phi(a)+\phi(b)$ and $\phi(ab) = \phi(a)\phi(b)$ for any pair a, b of elements.

irreducible polynomial: A polynomial with integer or rational coefficients is irreducible if and only if it cannot be factored as a product of two polynomials of strictly lower degree with rational coefficients.

limit of a sequence: Let x_n be a sequence. Formally, $\lim x_n = u$ if and only if, for each given positive real number ϵ, a number N (depending on ϵ) can be found for which $|x_n - u| < \epsilon$ whenever the index n exceeds N. More informally, x_n tends towards u as n increases iff x_n gets arbitrarily close to u as we take larger and larger values of n.

lowest terms: A common fraction is in lowest terms if its numerator and denominator are coprime.

matrix: A matrix is a rectangular array of numbers.

monic polynomial: A monic polynomial is a polynomial for which the coefficient of the highest power of the variable (leading coefficient) is equal to 1.

nonempty (nonvoid) set: A set is nonempty if it contains at least one element.

norm: The norm of a quadratic surd $a + b\sqrt{d}$ is the product of the surd with its surd conjugate, namely $a^2 - bd^2$. The norm of an algebraic number θ is the product of θ and all the other roots of an irreducible polynomial with integer coefficients with θ as a root.

parameter: A parameter in an equation is an algebraic quantity whose value may be one of a specific set of numbers, but which is regarded as being constant with respect to other variables in the equation.

parity: The parity of an integer refers to the characteristic of being even or odd.

pigeonhole principle: If you distribute n objects into m categories, and $n > m$, then there is a category that receives at least two objects.

polynomial: A polynomial is an expression of the form $a_nx^n + a_{n-1}x^{n-1} + \cdots + a_1x + a_0$ where n is a nonnegative integer, a_i are numbers (coefficients), $a_n \neq 0$ and x is a variable. The degree of the polynomial is n, the highest exponent of the variable.

quadratic polynomial: A quadratic polynomial is a polynomial of degree 2.

quadratic surd: A quadratic surd is a number of the form $a + b\sqrt{d}$ where a, b are rationals and d is an integer.

rational: A real number is rational iff it can be written in the form p/q where p and q are integers; each rational number can be written in *lowest terms*, for which the greatest common divisor of the numerator p and denominator q is equal to 1.

recursion: A recursion is a sequence $\{x_n\}$ which is defined by specifying a certain number of initial terms and then by giving some general rule by which each term is written as a function of its predecessors. An example is a *linear second order recursion* where the first two terms of the sequence are given and each subsequent term has the form $x_n = ax_{n-1} + bx_{n-2}$, where a and b are fixed multipliers. If $x_1 = x_2 = 1$ and the multipliers a and b are both 1, then we get the *Fibonacci sequence*.

ring: A subset of numbers is a ring if and only if it is closed under addition, subtraction and multiplication.

root of a polynomial: A root of a polynomial of a single variable is a number at which the polynomial takes the value zero. A multiple root r is one for which $(x - r)^n$ is a factor of the polynomial with variable x for some positive integer n exceeding 1.

root of unity: A complex number z is an nth root of unity if and only if it satisfies $z^n = 1$. Such a root is **primitive** iff it is not an mth root of unity with m less than n.

sequence: A sequence is an ordered set of numbers $\{x_n\}$, where n ranges over a set of consecutive integers, usually the positive or nonnegative integers, but sometimes the set of all integers. If the index n ranges over all integers, the sequence is said to be bilateral.

squarefree; An integer is squarefree iff it is not divisible by any square except 1.

surd conjugate: The surd conjugate of the quadratic surd $a+b\sqrt{d}$ is $a-b\sqrt{d}$.

upper bound; An upper bound of a set of numbers is a number which is at least as big as each number in the set. The least upper bound of the set of the smallest of the upper bounds.

References

A list of books is followed by a list of selected papers. For a general introduction to number theory, the books of **Friedberg**, **Burn**, **Leveque**, **Niven, Zuckerman and Montgomery**, **Hardy and Wright**, and **Hua** are recommended. The book by **Burn** is a collection of problems. For an introduction into the history, past and contemporary, of Pell's equation, consult the survey papers of **Lenstra** and **Williams**. The books by **Buelle** and **Davenport** provide an introduction to quadratic forms. Those interested in cubic problems can consult the book of **Delone and Faddeev**, as well as the papers of **Cusick**, **Daus**, **Mathews** and **Selmer**. For a more advanced treatment of algbraic number theory, see **Borevich and Shaferevich**, **Fröhlick and Taylor** and **Mollin**.

Books

A.K. Bag, *Mathematics in Ancient and Mediaeval India*. Varanassi, Chaukhambha Orientalia, 1979.

Z.I. Borevich and I.R. Shafarevich, *Number Theory*. Academic Press, 1966

Duncan A. Buelle, *Binary Quadratic Forms: Classical Theory and Modern Computation*. Springer, New York, 1989.

R.P. Burn, *A Pathway to Number Theory* (*second edition*). Cambridge, 1997.
Chapter 8: Quadratic forms
Chapter 10: Continued fractions (Pell's equation: pp. 223–226)

Henri Cohen, *A Course in Computational Number Theory*. Springer-Verlag, 1993.

Harold Davenport, *The Higher Arithmetic: An Introduction to the Theory of Numbers*. Hutchinson's University Library, London, 1952.
Chapter III: Quadratic residues
Chapter IV: Continued fractions (Pell's equation)
Chapter VI: Quadratic forms

Richard Dedekind, *Theory of Algebraic Integers*. Cambridge, 1996 (English translation of original French version of 1877).

Comment. Despite its age and the progress in the field since it was written, this book is still clear and well worth reading.

B.N. Delone and D.K. Faddeev, *The Theory of Irrationalities of the Third Degree.* Translations of Mathematical Monographs, American Mathematical Society, Providence, 1964.

Leonard Eugene Dickson, *History of the Theory of Numbers.* Chelsea, 1952, 1966.
Volume II: Chapter IV: Rational right triangles
Volume II: Chapter VII: Pell's equation
Volume III: Quadratic and higher forms

Leonard Eugene Dickson, *Introduction to the Theory of Numbers.* Chicago, 1929.
Chapter V: Binary quadratic forms
Chapter VI: Certain diophantine equations
Chapter VII: Indefinite binary quadratic forms (continued fractions)

Leonhard Euler, *Elements of Algebra=Opera Omnia* (1) 1, 1–498. (An English translation was published by Springer, New York, in 1984.)
Chapter 6: $ax^2 = b = \square$; Chapter 7: $ax^2 + 1 = \square$

David Fowler, *The Mathematics of Plato's Academy: A New Reconstruction. Second edition.* Oxford, 1999.

Richard Friedberg, *An Adventurer's Guide to Number Theory.* McGraw-Hill, New York, Toronto, London, Sydney, 1968.

A. Fröhlich and M.J. Taylor, *Algebraic Number Theory.* Cambridge, 1991.
Chapter IV: Units (Dirichlet unit theorem: section 4)
Chapter V: Fields of low degree (quadratic, biquadratic, cubic, and sextic)

Karl F. Gauss, *Disquisitiones arithmeticae. Transl. Arthur A. Clarke*, Springer, 1986.

A.O. Gelfond, *The Solution of Equations in Integers. Translated by J.B. Roberts*, Golden Gate, W.H. Freeman, 1961.

George Gheverghese Joseph, *The Crest of the Peacock: Non-European Roots of Mathematics.* I.B. Tauris, London and New York, 1991.

Anthony A. Gioia, *The Theory of Numbers.* Markham, Chicago, 1970.
Chapter 6: Continued fractions; Farey sequences; The Pell equation

Emil Grosswald, *Topics from the Theory of Numbers.* (*Second edition*). Birkhäuser, Boston, Basel, Stuttgart, 1984.
Chapter 10: Arithmetic number fields (Dirichlet unit theorem stated: section 10.13)

G.H. Hardy and E.M. Wright, *An Introduction to the Theory of Numbers. Fifth edition*, Oxford, 1979.
Chapter X: Continued fractions

L.K. Hua, *Introduction to Number Theory.* Springer-Verlag, Berlin, 1982.

H. Konen, *Geschichte der Gleichung* $t^2 - Du^2 = 1$. Leipzig, 1901.

William J. LeVeqeue, *Topics in Number Theory Volume I.* Addison-Wesley, Reading, MA.
Chapter 8: Pell's equation

William J. LeVeque, *Topics in Number Theory. Volume II.* Addison-Wesley, Reading, MA, 1956.
Chapter 2: Algebraic numbers (groups of units: section 2.9)

Neil H. McCoy, *The Theory of Numbers.* Macmillan, New York, 1965.
Chapter 5: Continued fractions

Richard A. Mollin, *Fundamental Number Theory with Applications.* CRC Press, 1998.
Chapters 5, 6

Richard A. Mollin, *Algebraic Number Theory.* CRC Press, 1999.

L.J. Mordell, *Diophantine Equations* Academic Press, 1969.
Chapter 8

Ivan Niven, Herbert S. Zuckermann, and Hugh L. Montgomery, *An Introduction to the Theory of Numbers. Fifth edition.* John Wiley & Sons, 1991.
Chapter 7: Continued fractions
Chapter 9: Algebraic numbers

Hans Rademacher, *Lectures in Elementary Number Theory.* Blaisdell, New York, 1964.

Paolo Ribenboim, *Algebraic Numbers.* Wiley-Interscience, 1972.
Chapter 9: Units and estimations for the discriminant (Dirichlet unit theorem: p. 148)

Kenneth H. Rosen, *Elementary Number Theory and its Applications. (Third edition).* Addison-Wesley, Reading, MA, 1993.
Chapter 10: Decimal fractions and continued fractions
Chapter 11: Some nonlinear diophantine equations (Pell's equation: section 11.4)

James E. Shockley, *Introduction to Number Theory.* Holt, Rinehart & Winston, 1967.
Chapter 12: Continued fractions and Pell's equation

David E. Smith, *ed., A source book in mathematics, Volume One.* Dover, 1959 (pages 214–216).

Ian Stewart and David Tall, *Algebraic Number Theory (Second edition).* Chapman & Hall, London, New York, 1987.
Chapter 12: Dirichlet's Unit Theorem
Comment. This is a very good book to start with for algebraic number theory.

D.J. Struik, *ed., A source book in mathematics, 1200–1800.* Harvard University Press, Cambridge, MA, 1969 (pages 29–31).

J.P. Uspensky and M.A. Heaslet, *Elementary Number Theory.* McGraw Hill, 1939.
Chapter XI: Some problems connected with quadratic forms

André Weil, *Number Theory: An Approach Through History from Hanmurapi to Legendre*. (Birkhäuser, 1984).

E.E. Whitford, *The Pell Equation*. College of the City of New York, New York, 1912.

Papers

Lionel Bapoungué, Un critère de résolution pour l'équation diophantienne $ax^2 + 2bxy - kay^2 = \pm 1$. *Expositiones Mathematicae* 16 (1998) 249–262.

J.M. Barbour, Music and ternary continued fractions. *Amer. Math. Monthly* 55 (1948) 545–555 **MR** 10 284.

Viggo Brun, Music and Euclidean algorithms. *Nordisk Mat. Tidskr.* 9 (1961) 29–36, 95 **MR** 24 (1962) A705.

Zhenfu Cao, The diophantine equations $x^4 - y^4 = z^p$ and $x^4 - 1 = dy^q$. *C.R. Math. Rep. Acad. Sci. Canada 21:1* (1999) 23–27.

A. Cayley, Note sur l'équation $x^2 - Dy^2 = \pm 4$, $D \equiv 5 \pmod 8$. *Journal für die reine und angewandte Mathematik* 53 (1857) 319–371.

C.C. Chen, A recursive solution to Pell's equation. *Bulletin of the Institute of Combinatorics and Applications* 13 (1995) 45–50.

T.W. Cusick, The Szekeres multidimensional continued fraction. *Mathematics of Computation* 31 (1977), 280–317.

T.W. Cusick, Finding fundamental units in cubic fields. *Math. Proc. Camb. Phil. Soc.* 92 (1982), 385–389.

T.W. Cusick, Finding fundamental units in totally real fields. *Math. Proc. Camb. Phil. Soc.* 96 (1984), 191–194.

T.W. Cusick and Lowell Schoenfeld, A table of fundamental pairs of units in totally real cubic fields. *Mathematics of Computation* 48 (1987) 147–158.

P.H. Daus, Normal ternary continued fraction expansion for the cube roots of integers. *Amer. J. Math.* 44 (1922) 279–296.

P.H. Daus, Normal ternary continued fraction expansions for cubic irrationals. *Amer. J. Math.* 51 (1929) 67–98.

Leonard Euler, De usu novi algorithmi in problemate pelliano solvendo. *Novi comm. acad. scientarium Petropolitanae* 11 (1765) 1767, 28–66=O.O. (1) 3, 73–111.

Leonard Euler, De solutione problematum diophanteorum per numeros integros. *Commentarii academiae scientarium Petropolitanae* 6 (1732/3) 1738, 175–188=O.O. (1) 2, 6–17.

Leonard Euler, Nova subsida pro resolutione formulae $axx + 1 = yy$. *Opscula analytica* 1 (1783) 310–328=O.O. (1) 4, 76–90.

M.J. Jacobson, Jr. and H.C. Williams, The size of the fundamental solutions of consecutive Pell equations. *Experimental Math* 9 (2000), 631–640.

J.L. Lagrange, Solution d'un problème d'arithmétique. *Misc. Taurinensia* 4 (1766–1769)=Oeuvres I, 671–731.

J.L. Lagrange, Sur la solution des problèmes indéterminés du second degré. *Mémoires de l'Académie Royale des Sciences et Belle-Lettres de Berlin* XXIII, 1769 = Oeuvres II, 379–535.

J.L. Lagrange, Additions aux Élements d'Algèbre d'Euler. Oeuvres VII, 5–180. Chapters VII, VIII, and pages 74–77.

J.L. Lagrange, Nouvelle méthode pour résoudre les problèmes indéterminés en nombres entiers. *Mémoires de l'Académie Royale des Sciences et Belle-Lettres de Berlin* XXIV, 1770 = Oeuvres II, 655–726.

D.H. Lehmer, A list of errors in tables of the Pell equation. *Bull. Amer. Math. Soc.* 32 (1926), 545–550.

D.H. Lehmer, On the indeterminate equation $t^2 - p^2Du^2 = 1$. *Annals of Mathematics* 27 (1926), 471–476.

D.H. Lehmer, On the multiple solutions of the Pell equation. *Annals of Mathematics* 30 (1928), 66–72.

H.W.Lenstra, Jr., Solving the Pell equation. *Notices of the American Mathematical Society* 49:2 (Feb., 2002), 182–192.

G.B. Mathews, On the arithmetic theory of the form $x^3 + ny^3 + n^2z^3 - 3nxyz$. *Proceedings of the London Mathematical Society* 21 (1980), 280–287.

K. Matthews, The Diophantine equation $x^2 - Dy^2 = N$, $D > 0$. *Expositiones Mathematicae* **18** (2000), 323--331.

E. Meissel, Betrag zur Pell'scher Gleichung höherer grad. *Proc. Kiel (1891).*

R.A. Mollin, All solutions of the Diophantine equation $x^2 - Dy^2 = n$. *Far East. J. Math. Sci.*, Special Vol. Part III, (1998), 257–293.

R.A. Mollin, Simple continued fraction solutions for Diophantine equations. *Expositiones mathematicae* 19 (2001), 55–73.

Ernst S. Selmer, Continued fractions in several dimensions. *Nordisk Mat. Tidskr.* 9 (1961) 37–43, 95 **MR** 24 (1962) A706.

C.-O. Selenius, Rationale of the chakravala process of Jayadeva and Bhaskara II. *Historia Mathematica* 2 (1975), 167–184.

Harry S. Vandiver, Application of a theorem regarding circulants. *American Mathematical Monthly* 9 (1902), 96–98.

Ilan Vardi, Archimedes' cattle problem. *American Mathematical Monthly* 105 (1998), 305–329.

W. Waterhouse, On the cattle problem of Archimedes. *Historia Mathematica* 22 (1995), 186–187.

H.C. Williams, R.A. German and C.R. Zarnka, Solution of the cattle problem of Archimedes *Math. Comp.* 19 (1965), 671–674.

H.C. Williams, Solving the Pell's equation *Number theory for the Millennium, Proc. Millennial Conf. Number Theory* (Urbana, IL) 2000 (M.A. Bennett *et al*, editors), A.K. Peters, Boston, 2002.

Index